Jay Miller

Presented by
The Fort Worth Museum of Science and History Association
June 29, 1991 – January 1, 1992

FIRST PRINTING

Soviet Space, an exhibition in the city of Fort Worth

Project Director: Robert H. Townsend
Project Coordinator: Kari Metroka
Design/Production: Dale Rushing, Nancy Draughn,
     Stan Slejko, *The Phillips Agency*
Editor: Jay Miller, *Aerofax, Inc.*
Consulting Editor: Stuart Feigley, *The Phillips Agency*
Contributing Editors: Harry Campbell, Kari Metroka, Laura L. Daly
Technical Editor: Melvin H. Schuetz
Illustration: Bill Dale, The Phillips Agency
Photography: Space Commerce Corporation,
     Private Collection of Jay Miller, *Aerofax, Inc.*, NASA,
     Exhibit Artifacts: Andy Post, Dallas
Timeline: Melvin H. Schuetz
Index: Pam Falejczyk-Crosby, *New Media Productions*
Color Separations: Classic Color
Printing: Williamson Printing Corporation
Typography: The Phillips Agency
Logo: Regian Advertising & Public Relations

Published in the United States in 1991 by the
Fort Worth Museum of Science and History Association

ISBN: 0-9629867-3-9
Library of Congress Catalog Card Number: 91-65880

© 1991 by The Fort Worth Museum of Science and History Association
All Rights Reserved

CONTENTS

Early Soviet Rocketry	1
Satellites and Planetary Probes	17
Terrestrial Satellites	21
Lunar Exploration Probes	29
Planetary Flights	37
Cosmonauts in Space	49
Space Stations	68
International Cooperation in Space	95
Soviet Space Program Today and Tomorrow	99
Suggestions for Further Reading	107
Index	108
Glossary	110

ABOUT THIS BOOK

Until recently, very little information concerning the Soviet space program and its equipment had been published by the U.S.S.R. Therefore, many of the documents used as sources in the writing of this book were from American experts knowledgeable in the space program of the Soviets. These documents, however, were never verified by Soviet officials. Every attempt has been made to include only information that is correct, as well as the current Soviet perspective as released by Soviet officials.

A MESSAGE FROM THE CO-CHAIRS, SOVIET SPACE ORGANIZING COMMITTEE

The Board of Trustees of the Fort Worth Museum of Science and History Association is honored to have you join us in celebrating the Museum's 50th Anniversary. From its beginning as the Fort Worth Children's Museum, the Museum's mission has been to provide a unique combination of quality science and history exhibits and programs through diverse educational experiences for every visitor.

Soviet Space embodies the Museum's commitment to its mission.

We are truly gratified for the support *Soviet Space* has received from the business and civic leaders and volunteers here in Fort Worth and throughout North Texas. Any undertaking of this size and scope must, ultimately, become a community effort, and *Soviet Space* certainly is such an effort.

In order to round out your experience and to enhance your understanding of the importance of space exploration, we also urge you to visit the Museum itself to view *Blue Planet* in the Omni Theater, *Space Spinoffs*, an exhibit that demonstrates the practical, everyday benefits of space research and development, and *Space Race*, a special space-oriented program in the Noble Planetarium. Together with *Soviet Space*, these Museum activities provide your family a fascinating introduction to space exploration and the world of the future.

Soviet Space is the Museum's birthday gift to you. We hope you will visit the Fort Worth Museum of Science and History again soon.

James L. Ervin
President
Soviet Space Co-Chair

Carol Beech
Vice President
Soviet Space Co-Chair

FORT WORTH MUSEUM OF SCIENCE AND HISTORY

Board of Trustees
J. Turner Almond
Anne Bass
Carol Beech
Gretchen Denny
Lisa Diffley
James L. Ervin
C. Reid Ferring
Preston M. Geren, Jr.
Richard A. Greenman
W. Elray Howard
Erma Johnson
Mark Johnson
Robert M. Lansford
Robert J. Mitchell
Christene Moss
Melvin E. Olsen
Ronald L. Parrish
Bill G. Prince
Paul R. Ray, Jr.
Reginald L. Robinson
Mauro Serrano
Gordan W. Smith
James R. Toal
Elaine Yamagata

James L. Ervin
President

Carol Beech
Vice President

Gretchen Denny
Secretary

Bill G. Prince
Treasurer

Anne Bass
Member-at-Large

Robert J. Mitchell
Member-at-Large

Ronald L. Parrish
Member-at-Large

SOVIET SPACE ORGANIZING COMMITTEE

Co-Chairs
James L. Ervin
Carol Beech

Marketing/Public Relations
Ronald L. Parrish
Jane Schlansker

Gala
Pat Massad

Speakers
Christene Moss

Community Advisory
Erma Johnson
John J. Hernandez
Devoyd Jennings
Terrance J. Ryan

Finance/Accounting
J. Turner Almond
Judy Hunter

Sponsorship/Development
Paul R. Ray, Jr.
Mark Hart, Jr.

Hospitality
Elaine Yamagata
Douglas Harman

Education/Outreach
Mauro Serrano
Dr. John McElroy

Volunteer
Gretchen Denny
Elizabeth Ray

A MESSAGE FROM THE GOVERNOR OF TEXAS

On behalf of the people of Texas, I am delighted to send greetings to our distinguished visitors from the Soviet Union who are part of *Soviet Space*, this extraordinary exhibit celebrating the 50th Anniversary of the Fort Worth Museum of Science and History. *Soviet Space* presents an unprecedented look at the accomplishments of the Soviet Union's impressive space program, and it is my hope that this educational exchange will foster better international understanding and cooperation.

Additionally, I want to salute the many people in Fort Worth and the Metroplex who have made *Soviet Space* a reality. The educational and cultural impact of *Soviet Space*, not to mention its significant economic impact on the State, are sources of great pride for all of us. I commend the Board of the Fort Worth Museum of Science and History for its visionary leadership in making *Soviet Space* possible. A hearty "Thank You" also goes to the Fort Worth community leaders, the corporate sponsors and the hundreds of volunteers throughout North Texas who have helped make *Soviet Space* a resounding success.

It is also a great privilege to welcome the thousands of visitors to our State who are attending this magnificent exhibit. We hope you enjoy *Soviet Space* and your stay in Texas.

Ann Richards
Governor of Texas

A MESSAGE FROM THE MAYOR OF FORT WORTH

As Mayor of the City of Fort Worth, it is a distinct pleasure to commend the Board of the Fort Worth Museum of Science and History for bringing *Soviet Space* to our city. Throughout its 50 years as a leading educational and cultural institution, the Museum of Science and History has brought distinction to Fort Worth; the Museum and its Omni Theater are among the most popular entertainment and educational attractions in North Texas, annually drawing more than one million visitors each year. I am confident *Soviet Space* will launch another half century of greatness for the Museum and for our community.

    This outstanding exhibit truly has brought our community together, and I personally want to thank each individual and corporate sponsor who has helped make this project an unparalleled success.

    Fort Worth is greatly honored to be the exclusive Southwestern venue for this spectacular collection of items from the Soviet space program. For those of you who are visitors to our city, we hope you enjoy *Soviet Space* as well as your stay in our city. While you are here, please take advantage of the opportunity to see *Blue Planet*, *Space Spinoffs* and *Space Race* at the Museum of Science and History. Other attractions in our city include the Kimbell Museum, the Amon Carter Museum of Western Art and the Modern Art Museum, all within easy walking distance of *Soviet Space*. We hope you'll also take this opportunity to enjoy the rich heritage of our historic Stockyards area and beautiful downtown.

    Soviet Space is a marvelous gift to Fort Worth, the Fort Worth-Dallas communities, Texas and the Southwest.

Kay Granger
Mayor of Fort Worth

A MESSAGE FROM THE SOVIET AMBASSADOR

I would like to welcome you on the occasion of the *Soviet Space* exhibit in the Fort Worth Museum of Science and History.

With the 1957 launch of *Sputnik*, the first artificial satellite, the era of Soviet space was prepared for take-off. Thirty years ago, on April 12, 1961, Yuri Gagarin became the first man on the Earth who had flown over our planet on board the Soviet *Vostok* spaceship.

In the following decades, my country put the first woman in space, the first man to walk in outer space and carried out the total of 70 manned missions from Baikonur to the Earth orbit. The space flight time by all Soviet cosmonauts has exceeded some 7,400 man-days or over 20 man-years.

The Soviet Union was involved in several international space projects. In 1975 there was the famous *Soyuz/Apollo* mission, which provided a good example that experts from the two countries can cooperate fruitfully and to mutual benefit. Identical human ingenuity and courage were tested. This significant mission underscored the fact that we can seek common goals being motivated by a common desire for peace.

The world looks up to the space programs of the Soviet Union and the United States. The Fort Worth Museum of Science and History now gives the world the opportunity to view space on Earth. Americans can experience genuine American/Soviet efforts to work together. In space, cooperation for the benefit of mankind is a magnificent goal which may be achieved.

It is often said that the universal language is music. The universe, though, provides our countries with a common space for communication and at the same time, a space for pursuing what we seek to accomplish together on Earth.

Viktor Komplektov
U.S.S.R. Ambassador to the United States

A MESSAGE FROM THE CHAIRMAN OF GLAVKOSMOS

The end of the Cold War along with the advent of *glasnost* has given us the opportunity to share a part of our space program with the United States of America.

The space programs of the United States and the Soviet Union have achieved great victories as well as suffered tragic defeats. Both Russians and Americans share the same aspirations and challenges in the wondrous world of space. Both have great dreams and high hopes.

As we approach the turn of the century, it is important that our efforts in space become shared efforts. The expertise of one country can benefit the other. The more we work together, the more we will grow to understand each other. After all, we share one planet as we do one universe. In space exploration, not even the sky is our limit. If we work together, the possibilities are boundless.

It is with these thoughts that I welcome the exhibit *Soviet Space* in Fort Worth. May this exhibit spur dreams among American children and adults alike. Let us share the adventure in space together. If we have not yet met on this Earth, let us meet in space.

Alexander Dunaev

Chairman, GLAVKOSMOS

The Central Administration of Space Technology Development and Use for the

National Economy and Science

SOVIET HONORARY COMMITTEE

Anatoly M. Baklunov
General Director, NPO Lavochkin

Vladimir P. Barmin
Academician, General Designer
Ministry of General Machine Building

Anatoly A. Chizhov
General Director, Progress Plant (Samara)

Alexander I. Dunaev
Chairman, GLAVKOSMOS

Boris P. Katorgin
General Designer, NPO Energomash

Yuri N. Koptev
Vice Minister, Ministry of General Machine Building

Gay I. Severin
General Director, Zvezda Plant

Vladimir A. Shatalov
Lieutenant General, Commander, Cosmonaut Training Center

Yuri M. Solomko
Director, Memorial Museum of Cosmonautics

SOVIET OFFICIAL DELEGATION

Ignat I. Askerko
Chief Spacesuit Expert, Zvezda Plant

Stanislav P. Bogdanovskiy
Head of NPO Energomash

Arthur M. Dula
President, Space Commerce Corporation

Konstantin P. Feoktistov
Pilot-Cosmonaut

Oleg Y. Firsyuk
Vice Chairman, International Affairs, GLAVKOSMOS

Alexander Kirilin
Deputy Director, Progress Plant

Sergei V. Kovalevskiy
Director, Znanie Plant

Alexei P. Milovanov
Chief Manager, Soviet Aerospace

Sergei N. Mironov
Expert, GLAVKOSMOS

Dimitry Poletaev
Head of Department, GLAVKOSMOS

Alexander L. Rodin
Deputy General Designer, NPO Lavochkin

Svetlana Y. Savitskaya
Pilot-Cosmonaut

Andrei P. Sergeyuk
Expert, GLAVKOSMOS

Vadim V. Shultsev
Supervising Manager, Ministry of General Machine Building

SOVIET SPACE SPONSORS

Major Sponsors
Amon G. Carter Foundation
Dallas Times Herald
Fort Worth Star-Telegram
KPLX 99.5/KLIF 570
KXAS - TV
NCNB Texas
Regian Advertising &
    Public Relations
Tandy Corporation
WFAA - TV

Underwriters
Perot Foundation
Sid W. Richardson
    Foundation

Official Airline
Delta Air Lines

Sponsors
Mr. & Mrs. Robert M. Bass
Dallas Morning News
Adeline and George McQueen
    Foundation/TeamBank,
    Trustee
PR/Texas
Rainbow - TicketMaster/
    TicketQuik
Paul Ray & Carré Orban
    International
Prime Time Video Inc.
Union Pacific Foundation/
    Union Pacific Resources
    Company
The Worthington Hotel

Patrons
Burlington Northern
    Foundation representing
    Burlington Northern
    Railroad
Chaparral Steel
Exhibitgroup Inc.
Garvey Texas Foundation, Inc.
Junior League of
    Fort Worth, Inc.
KTVT Channel 11
The Phillips Agency
Simplex Time Recorder
Texas Monthly
TU Electric
Williamson Printing Corporation

Donors
Color Tile, Inc.
Crystelle Waggoner
    Charitable Trust
Cyrix Corporation
Miller Brewing Company,
    Fort Worth
William E. Scott Foundation
Stevens Graphics
Source Telecommunications
White Decorating Service

Friends
Alcon Foundation
Jack Daugherty Fund of the
    Community Foundation of
    Metropolitan Tarrant Co.
Coopers & Lybrand
Coors Distributing Co.
General Dynamics Corporation,
    Fort Worth Division
Kelly, Hart & Hallman
Law, Snakard & Gambill
    Attorneys and Counselors
Luther King Capital
    Management

Friends
Mary Potishman Lard Trust
McDonald's
Murata Business Systems, Inc.
National Guardian Security
    Services Corporation
NTS, Inc.
Overton Park Bank
Shannon, Gracey, Ratliff & Miller
Southwestern Bell Foundation
Williamson-Dickie Manufacturing
    Company
Anonymous

A MESSAGE FROM THE EXECUTIVE DIRECTOR OF THE FORT WORTH MUSEUM OF SCIENCE AND HISTORY

Soviet Space marks a milestone for the Fort Worth Museum of Science and History. Never before in our fifty years as an educational and cultural institution have we offered such a comprehensive and complex exhibition program. We believe this dynamic presentation, celebrating the Museum's 50th Anniversary, will broaden your horizons and challenge your limits just as it has the Museum's.

The Museum staff has long envisioned bringing an exhibit of this magnitude to our community, and I thank the members of the Museum Board of Trustees for sharing that vision and making it a reality. I also wish to express our sincerest thanks to our donors and corporate sponsors for providing the resources necessary to make *Soviet Space* possible.

We are proud to offer our visitors and friends a first glimpse into the mysterious and obscure world of GLAVKOSMOS and the Soviets' remarkable space achievements.

Donald R. Otto

Executive Director
Fort Worth Museum of Science and History

FORT WORTH MUSEUM OF SCIENCE AND HISTORY STAFF

Donald R. Otto
Executive Director

Associate Directors
Maria Hall
Administration

Linda K. Johnson
Programs

Charlie Walter
Interpretation

Accounting/Personnel
Lynn Hamlett
Lisa Miller
Mary Mahanay
Sharon Harris
Jean Manning
Donna Scaroleta

Development
Judie B. Greenman
Carol P. Hendrix
Holly Henry
J.B. McCauley
Alison Ashmore

Education
Kit Goolsby
Miki Gabbard
Barbara Maguire
Colleen Blair

Executive Administration
Jeanne Boyd
Karla Berry

Exhibits
Derek Taulman
Chris Hailey
Dennis Gabbard
Robert Johnson

History
Terry Grose
Renee Erwin
Jane Dees
Joyce Williams

Maintenance
Dale Wrinkle
Tiffany Ray
Jerry Martin
David McCauley
Frank Bauer
George Batty
Joe Amezcua
Charles Bates
Geno Contreras
Delores Cook
Johnnie Hafford
Johnnie Mae Harris
Mae Houston
Christene James
Maria Lukacs
Gary Scott
Johnny Thompson
Billy Williams
Brenda Wyatt

Museum Store
Anne Garrison
Merla Kramer
Jennifer Childs

Production Services
Don Garland

Public Relations
Karen Turner
Erin Offill

Science
William J. Voss
James P. Diffily
Wesley Hathaway
Robert L. Lindsey

Security
Fitz Maguire
Ed Carson
Bobby Hayes
Christopher Hailey

Theater
Richard Van Zandt
Robert Esterlein
Jane Dees

Visitor Services
Charlie Walter
Betsy Cummings
Charlene Ingram
Daniel Connor
Nita Overton
Monica Berry
Karen Bohling

Volunteer Services
Karen Turner
Joyce Williams

SOVIET SPACE STAFF

Robert H. Townsend
Executive Director

Administration

Patricia G. Smith
Finance and Accounting Manager

Joan Mayo
Administrative Assistant

Kayla Epping
Administrative Assistant

Paula Wood
Receptionist

Operations

Tim Gette
Operations Director

Richard Frink
Senior Floor Manager

Mark Vinson
Senior Floor Manager/Interpreter

Randall Peters
Floor Manager

Kathy Ferguson
Gift Shop Manager

Karen Andrews
Nancy Wilson
Assistant Gift Shop Managers

Laura Barber
Gift Shop Assistant

Marketing/Public Relations

Harry Campbell
Marketing/Public Relations Director

Betsy Barnes
Special Events Manager

Chris Joubert
Group Sales Marketing Manager

Deborah Leliaert
Public Relations Manager

Kari Metroka
Marketing Manager

Lisa Newsom
Marketing Coordinator

Education/Special Programs

Judy Rufner
Education/Special Programs Director

Laura L. Daly
Camp-In Manager

Melvin H. Schuetz
Education Research Manager

Cheryl Stewart
Education Coordinator

Volunteers

Cathy Moates
Volunteer Director

Nancy Hankamer
Ann Bastable
Irma Perez
Linda Wood
Volunteer Coordinators

Ticketing Services

Brad Oldham
Group Sales Manager

Penny Orona
Box Office Manager

Laura Couch
Alisa Maples
Assistant Ticketing Managers

Lynn Alexander
Kay Anderson
Lisa Arbuckle
Elizabeth Clifford
Arlene Thompson
Group Ticket Sales Representatives

ACKNOWLEDGEMENTS

The astonishing success of the Soviet Space program, like all challenging and innovative projects, depends on the active participation and support of many. So too, the production, staging and hosting of a blockbuster exhibition the likes of *Soviet Space* depends on ideas and assistance of individuals from around the community. Its creation is the dream of many U.S. space enthusiasts and the brainchild of the Museum of Science, Boston and GLAVKOSMOS, the Soviet civil space agency. Space Commerce Corporation, Houston, Texas and in particular, its President, Art Dula, played a critical role in facilitating this venture.

Fort Worth's involvement in this unique opportunity began with the visions of those at the Fort Worth Museum of Science and History, especially its Executive Director, Don Otto. The Museum Staff, led purposefully by Linda Johnson, and the Museum's Board of Trustees, under the leadership of Jim Ervin and Carol Beech, thoroughly and critically reviewed the Exhibition. Enthusiastically, they agreed to host *Soviet Space*, the largest exhibit ever to visit Fort Worth, in celebration of the Museum's 50th Anniversary. Soon thereafter came the wholehearted endorsement of a host of community leaders and sponsors who lent the early spiritual and financial support that allowed the Exhibition to become a reality. Mayor Bob Bolen, his City Council including current Mayor Kay Granger, leaders of the Fort Worth, Hispanic and Black Chambers of Commerce, and Doug Harman at the Convention and Visitors Bureau were particularly active in gaining support for the Museum's effort during this early period.

Because GLAVKOSMOS allowed the Fort Worth Museum of Science and History to completely reproduce the *Soviet Space* Exhibit, only our imaginations limited the way we presented this first Western view of the Soviet space program. We are grateful to those who allowed our ideas to run free, and in combination with their own, developed an Exhibition that will impress the greatest skeptic and entertain the most weary. Most particularly our gratitude for this exercise goes to David Gibson and his associates, Bill Hill and John Gerrits at Exhibitgroup Inc., for their innovative exhibit design; Terry and Sylvia McCullough for their attention-commanding audio/video expertise; Marshall Riggin for his thought-provoking script writing; Michael McLoughlin for his thorough audio guide production. All of these artists have made the Fort Worth venue of *Soviet Space* truly an unforgettable experience.

For those exhibition visitors who want to carry home the *Soviet Space* experience, this official Catalog of the Exhibit has been produced. In presenting a complete retrospective of the Soviet space program, we became indebted to those who shared their knowledge and opened their files for our review including: Michael Binder, Irwin Bulban, Douglas Champlin, Tom Copeland, Art Dula, Reuben Johnson, Chris Pocock, Albert Ross, and Melvin Schuetz. Thoughtful interpretation of difficult subject matter was provided by Phillip Clark, Martin Horwitz, Michelle Moore and David Woods. Our sincere appreciation goes to the catalog's author, Jay Miller, who took a very complex, voluminous and mysterious subject matter, and put it into a concise and readable form for novices and experts alike to enjoy. His moral and production supporters Anna, Miriam and Susan Miller ensured that tight time schedules were met. To all those at The Phillips Agency involved in its production, including Marsha Alvey, Christie Barrows, Stuart Feigley, Ann Delbridge and Jane Russell, we owe special thanks; and, especially to Randy Phillips, Dale Rushing, Stan Slejko and Nancy Draughn who collectively have made this catalog as artistically pleasing as it is informative. Unique images which bring the catalog to life are the product of Andy Post and Bill Dale. The richness of color reproduction and attractive appearance is due, in large measure, to the uncompromising standards of Jerry Williamson and his staffs at Williamson Printing and Classic Color. All of these individuals, under the meticulous direction of Kari Metroka, can be proud of turning out a quality product that even the most avid space enthusiast will want in his library.

Finally, the single most impressive contribution that makes a project of this magnitude successful is the human element which is abundantly apparent in every aspect of the Exhibition. Whether it is the community leaders who lent their expertise as chairpersons or members of the organizing committees, the thousands of volunteers who dedicated countless hours manning various posts in the Exhibit, the Exhibit staff who took a year out of their careers to accept the challenge of managing a once-in-a-lifetime project, or those other professionals like Julie Wilson, Tom Eddleman and Jeff Coleman at Regian Advertising, who along with Jane Schlansker and David Lindsey at PR/Texas, directed the magnificent advertising and public relations efforts for our Exhibit, they all deserve our utmost respect and outright thanks. Without them, *Soviet Space* would not have happened.

Robert H. Townsend

TIMELINE

Includes major space exploration firsts, events which set new manned space endurance records and other significant space related events.

1957 (USSR)
Sputnik 1 – World's first artificial satellite
Sputnik 2 – First animal in orbit (dog, Laika)

1957 (USA)
Vanguard TV-3 – First U.S. satellite launch attempt {failure}

1958 (USSR)
Sputnik 3 – First orbiting geophysical laboratory

1958 (USA)
Explorer 1 – First U.S. artificial satellite; discovered Van Allen radiation belts
Vanguard 1 – First satellite equipped with solar cells
National Aeronautics and Space Administration {NASA} established

1959 (USSR)
Luna 1 – First spacecraft to achieve Earth-escape velocity
Luna 2 – First spacecraft to impact the Moon
Luna 3 – First spacecraft to photograph the far side of the Moon

1959 (USA)
Vanguard 2 – First photographs transmitted by television
Jupiter – First U.S. primates in space; sub-orbital flight (monkeys Able and Baker)
Vanguard 3 – First spacecraft to map the Earth's magnetic field

1960 (USSR)
Sputnik 5 {*Korabl-Sputnik* 2} – First recovery of animals from orbital flight (2 dogs and 6 mice)

1960 (USA)
Tiros 1 – First meteorological satellite
Echo 1 – First communications satellite {experimental}

1961 (USSR)
Vostok 1 – First man in space (Gagarin)
Vostok 2 – First man to spend a day in space (Titov)

1961 (USA)
MR-3 *Freedom* 7 – First U.S. man in space; sub-orbital flight (Shepard)
President Kennedy announces national goal of a manned lunar landing by the end of the decade

1962 (USSR)
Vostok 3/*Vostok* 4 – First dual space mission; *Vostok* 4 passes within 4 mi [6.5 km] of *Vostok* 3 (Nikolayev) (Popovich)

1962 (USA)
MR-3 *Friendship* 7 – First U.S. manned orbital flight (Glenn)
Telstar 1 – First commercial communications satellite
Mariner 2 – First successful fly-by of Venus

1963 (USSR)
Vostok 5 – Longest solo space flight ever: 4 days, 23 hours, 6 minutes (Bykovsky)
Vostok 6 – First woman in space (Tereshkova)

1963 (USA)
MA-9 *Faith* 7 – First U.S. manned flight to exceed 24 hours (Cooper)

1964 (USSR)
Voskhod 1 – First 3-person orbital flight (Komarov, Yegorov, Feoktistov)

1964 (USA)
Ranger 7 – First close-up pictures of the Moon

1965 (USSR)
Voskhod 2 – First extra-vehicular activity {space walk, Alexei Leonov} (Belyayev, Leonov)
Proton 1 – First cosmic ray measurements

1965 (USA)
Gemini 3 – First U.S. 2-man space flight; first piloted orbital maneuvers (Grissom, Young)
Gemini 4 – First U.S. extra-vehicular activity {space walk, Ed White} (McDivitt, White)
Mariner 4 – First successful fly-by of Mars, first close-up pictures of the planet
Gemini 7 – Longest manned space flight to date; 13 days, 18 hours (Borman, Lovell)
Gemini 6 – First piloted space rendezvous {with *Gemini* 7} (Schirra, Stafford)

1966 (USSR)
Luna 9 – First unmanned landing on the Moon; first pictures from the lunar surface
Luna 10 – First unmanned spacecraft to orbit the Moon
Luna 13 – Unmanned lunar landing; first to test the density of lunar soil

1966 (USA)
Gemini 8 – First docking between manned and unmanned spacecraft {with *Agena* 8} (Armstrong, Scott)
Gemini 10 – First dual rendezvous in space {with unmanned *Agena* 10, then *Agena* 8} (Young, Collins)
Lunar Orbiter 1 – First pictures of Moon from lunar orbit
Gemini 11 – New altitude record set for manned space flight 850 mi [1368 km] (Conrad, Gordon)

1967 (USSR)
Soyuz 1 – First flight test of *Soyuz* ends tragically when parachute lines become entangled upon re-entry; (Vladimir Komarov) is killed on ground impact
Venera 4 – First successful Venus atmospheric probe
Kosmos 186,188 – First unmanned spacecraft to automatically rendezvous and dock in orbit

1967 (USA)
AS-204 *Apollo* 1 – Fire engulfs *Apollo* spacecraft during a launch pad rehearsal; (Grissom, White and Chaffee) are killed
Apollo 4 – First launch of *Saturn* V booster, with unmanned Command Module

1968 (USSR)
(Gagarin), the first man into space, is killed in a crash of his Mig-15 military jet
Zond 5 – First circumlunar flight and recovery of live animals (turtles)

1968 (USA)
Apollo 7 – First manned flight of *Apollo* program; first U.S. 3-man space flight (Schirra, Eisele, Cunningham)
Apollo 8 – First manned orbit of the Moon and return to Earth (Borman, Lovell, Anders)

1969 (USSR)
Soyuz 4/*Soyuz* 5 – First docking of two manned spacecraft and crew transfer (Shatalov, Khrunov, Yeliseyev, Volynov)
Soyuz 6 – First welding and smelting experiments in space (Shonin, Kubasov)

1969 (USA)
Apollo 10 – First lunar orbit rendezvous and docking (Stafford, Young, Cernan)
Apollo 11 – First manned landing on Moon and return, with 48.5 lb [22 kg] of lunar samples (Armstrong, Collins, Aldrin)

1970 (USSR)
Soyuz 9 – Longest manned space flight to date; 17 days, 16 hours, 59 minutes (Nikolayev, Sevastyanov)
Luna 16 – First automated lunar landing and return of soil samples
Luna 17 – First automated lunar roving vehicle, *Lunokhod* 1
Venera 7 – First successful landing on Venus

1970 (USA)
Apollo 13 – Third lunar landing attempt aborted due to explosion of oxygen tank in Service Module (Lovell, Swigert, Haise)

1971 (USSR)
Salyut 1 – World's first space station is orbited
Soyuz 11 – First crew to enter *Salyut*, all are killed during re-entry when cabin atmosphere is lost (Dobrovolsky, V. Volkov, Patsayev)
Mars 3 – First confirmed landing on Mars; lander failed after touch down due to dust storm

1971 (USA)
Apollo 15 – First use of Lunar Roving Vehicle on Moon (Scott, Worden, Irwin)
Mariner 9 – First successful Mars orbiter; first pictures of Mars from orbit

1972 (USSR)
In Moscow, Premier Kosygin and President Nixon sign space cooperation agreement and lay foundation for ASTP mission of 1975
Venera 8 – First analysis of soil on Venus

1972 (USA)
Apollo 16 – Deployment of first lunar astronomical observatory and first sub-satellite in lunar orbit (Young, Mattingly, Duke)
Apollo 17 – Last manned lunar landing, crew spends a record 74 hours, 59 minutes on Moon (Cernan, Evans, Schmitt)

1973 (USSR)
Soyuz 12 – Spacecraft requalification; first flight after *Soyuz* 11 failure (Lazarev, Makarov)

1973 (USA)
Skylab 1 – U.S. space station launched, is damaged during ascent
Explorer 49 – First radiotelescope in lunar orbit
Skylab IV – Final crew to occupy *Skylab*; record stay in space of 84 days, 1 hour, 15 minutes (Carr, Gibson, Pogue)
Pioneer 10 – First fly-by of Jupiter; first close-up pictures of the planet

1974 (USSR)
Salyut 3 – First operational military space station

1974 (USA)
Mariner 10 – First close-up pictures of Venus, followed by first fly-by and first close-up pictures of Mercury

1975 (USSR)
Soyuz 18A – First space flight aborted due to launch vehicle malfunction; crew survives (Lazarev, Makarov)
Soyuz 19 – First international docking, with *Apollo* {ASTP} (Leonov, Kubasov)

1975 (USA)
Apollo {ASTP} – First international docking, with *Soyuz* 19 (Stafford, Slayton, Brand)

1976 (USSR)
Luna 24 – First automated deep soil sample return from the Moon

1976 (USA)
Viking 1 – First pictures from the surface of Mars

1977
Salyut 6 – Launch of second generation Soviet space station

Soyuz 26 – Docked with *Salyut* 6, established new space endurance record of 96 days, 10 hours (Romanenko, Grechko)

1978
Soyuz 27 – First 3-spacecraft complex {with *Soyuz* 26 and *Salyut* 6} (Dzhanibekov, Makarov)

Progress 1 – First automated cargo spacecraft

Soyuz 28 – Docked with *Salyut* 6, first international crew in space (Gubarev, Remek)

Soyuz 29 – Docked with *Salyut* 6, established new space endurance record of 139 days, 14 hours, 48 minutes (Kovalenok, Ivanchenkov)

1979
Soyuz 32 – Docked with *Salyut* 6, established new space endurance record of 175 days, 36 minutes (Lyakhov, Ryumin)

Soyuz T-1 – Unmanned test of redesigned *Soyuz* spacecraft, which can carry three cosmonauts

1980
Soyuz 35 – Docked with *Salyut* 6, established new space endurance record of 184 days, 20 hours, 12 minutes (Popov, Ryumin)

Soyuz 38 – Cuban cosmonaut Arnaldo Tamayo Mendez became the first Hispanic in space (Tamayo, Romanenko)

1981
Soyuz T-4 – Docked with *Salyut* 6; flight engineer Savinykh becomes the 100th man in space (Kovalenok, Savinykh)

1982
Venera 13 – Lands on Venus, returns first color pictures of the surface

Salyut 7 – New space station is orbited, features more amenities for crew and improved docking unit

Soyuz T-5 – Docked with *Salyut* 7; established new space endurance record of 211 days, 8 hours, 5 minutes (Berezovoi, Lebedev)

Soyuz T-7 – Second woman in space, Savitskaya (Popov, Serebrov, Savitskaya)

1983
Soyuz T-9 – First construction in space {solar panel installation} (Lyakhov, Alexandrov)

Soyuz T-10A – First manned booster explosion, on launch pad; crew survives as emergency rocket escape system pulls them to safety (Strekalov, Titov)

1984
Soyuz T-10B – Docked with *Salyut* 7; V. Solovyov and Atkov established new space endurance record of 236 days, 22 hours, 50 minutes (Kizim, V. Solovyov, Atkov)

Soyuz T-12 – Docked with *Salyut* 7; first female to walk in space, Savitskaya (Dzhanibekov, Savitskaya, Volk)

The twin *Vega* probes are launched on their journeys to Venus and Halley's Comet

Kosmos 1614 – Last orbital flight of sub-scale model spaceplane as part of Soviet shuttle development

1977
First "approach and landing" test of space shuttle Orbiter *Enterprise*

The two *Voyager* probes are launched on their journeys to the outer solar system

1978
Pioneer-Venus 1 – First spacecraft to orbit Venus

1979
Voyager 1 – Fly-by of Jupiter; discovers ring around the planet and volcanos on the moon Io

Voyager 2 – Fly-by of Jupiter; discovers three new satellites

Pioneer 11 – First fly-by of Saturn; first close-up pictures of the planet

1980
Voyager 1 – Fly-by of Saturn; discovered {confirmed} six new moons and new details in rings

1981
STS-1 – First flight of a reusable spacecraft; orbital test flight of space shuttle *Columbia* (Young, Crippen)

Voyager 2 – Fly-by of Saturn; discovers four new moons

1982
STS-4 – Final test flight of space transportation system {shuttle}, future flights will be operational (Mattingly, Hartsfield)

STS-5 – First 4-person space flight (Brand, Overmeyer, Allen, Lenoir)

1983
Pioneer 10 – Passes the orbit of Neptune, becomes the first probe to leave the solar system

STS-7 – First U.S. woman in space, Ride; first 5-person space flight (Crippen, Hauck, Fabian, Ride, Thagard)

STS-8 – First African-American in space, Dr. Guion Bluford (Truly, Brandenstein, Gardner, Bluford, Thornton)

STS-9 – First 6-person space flight (Young, Shaw, Garriott, Parker, Lichtenberg, Merbold)

1984
STS-41B – First untethered space walk {first human satellite}, McCandless (Brand, Gibson, McNair, Stewart, McCandless)

STS-41C – First in-orbit satellite repair; Van Hoften, Nelson, (Crippen, Hart, Nelson, Scobee, Van Hoften)

STS-41G – First 7-person space flight (Crippen, Garneau, Leestma, McBride, Ride, Scully-Power, Sullivan)

STS-51A – First satellite retrieval and return from space; Gardner, Allen (Allen, Fisher, Gardner, Hauck, Walker)

1985
The twin *Vega* probes, en route to Halley's Comet, deploy landers and atmospheric balloons at Venus

Soyuz T-14 – First partial crew rotation {with *Soyuz* T-13 at *Salyut* 7} (Vasyutin, Grechko, A. Volkov, Dzhanibekov, Savinykh)

GLAVKOSMOS – Soviet civil space agency is established

1986
Mir – Launch of world's first modular space station

The twin *Vega* probes fly-by Halley's Comet returning data and pictures to Earth

Soyuz T-15 – First transfer between space stations {from *Mir* to *Salyut* 7 and back to *Mir*} (Kizim, V. Solovyov)

Soyuz TM-1 – First flight of improved *Soyuz* spacecraft {unmanned}

1987
Soyuz TM-2 – Docked with *Mir*; Romanenko established new space endurance record of 326 days, 11 hours, 38 minutes (Romanenko, Laveikin)

Kvant – New large astrophysics module is launched and docked with *Mir*

Energiya – First test flight of new heavy-lift launch vehicle; first Soviet vehicle utilizing liquid hydrogen fuel

Soyuz TM-4 – Docked with *Mir* complex; Manarov and Titov established new space endurance record of 365 days, 22 hours, 39 minutes (Titov, Manarov, Levchenko)

1988
The twin *Phobos* probes are launched on their journeys to Mars and its largest moon

Phobos 1 – Spacecraft is lost after faulty commands are sent by mission control

Buran – First test flight of Soviet space shuttle, launched by *Energiya* booster; (unmanned)

1989
Phobos 2 – Spacecraft is lost due to on-board computer failure; probe had reached Mars and was imaging the planet and its largest moon

Granat – Launch of large orbiting astrophysical observatory

Kvant 2 – Second large space station module is launched and docked with *Mir*

1990
Ikarus – First test in space of Soviet Manned Maneuvering Unit, the so-called "space motorcycle"

Progress 42 – Last old-style cargo supply craft is launched to *Mir*, new generation craft is designated *Progress*-M

Kristall – Third large space station module is launched and docked with *Mir*

Soyuz TM-11 – First journalist in space, Japanese reporter Toyohiro Akiyama, accompanies two Soviet cosmonauts to *Mir*; a TV station paid $12 million for the flight (Akiyama, Afanasyev, Manarov)

1991
Salyut 7 – Space station, orbited in 1982, ends its life, re-enters the atmosphere over South America

Informator – First of new series of satellites is launched for the U.S.S.R. Ministry of Geology

Almaz – Largest operational Earth resources satellite ever is orbited

1985
ICE – NASA's International Cometary Explorer is the first probe to reach a comet (Giacobini-Zinner)

STS-61A – First 8-person space flight (Hartsfield, Nagel, Bluford, Buchli, Dunbar, Furrer, Messerschmid, Ockels)

1986
Voyager 2 – First fly-by of Uranus; first close-up pictures of the planet

STS-51L – Space shuttle *Challenger* explodes 73 seconds after launch; all seven crew members are killed (Scobee, Smith, Resnik, Onizuka, Jarvis, McAuliffe, McNair)

1987

1988
STS-26 – First U.S. manned space flight since *Challenger* explosion in 1986 (Hauck, Covey, Lounge, Hilmers, Nelson)

1989
Magellan – Deployment of Venus radar mapper by shuttle mission *STS*-30

Voyager 2 – First fly-by of Neptune; first close-up pictures of the planet

Galileo – Deployment of Jupiter probe by shuttle mission *STS*-34

1990
LDEF – Long Duration Exposure Facility, in orbit since 1984 is retrieved and returned to Earth by shuttle mission *STS*-32

Galileo – On its way to Jupiter, makes fly-by of Venus

HST – Hubble Space Telescope is deployed by shuttle mission *STS*-31; mirror is subsequently discovered to be flawed

Magellan – Reaches Venus and begins mapping mission lasting 243 Earth days

Ulysses – Deployment of solar polar orbiter by shuttle mission *STS*-41

1991
Titan 4 – First launch of this new booster occurs at Vandenberg A.F.B.

GRO – Gamma Ray Observatory, large astrophysical satellite, is deployed by shuttle mission *STS*-37

Endeavour – New space shuttle, replacement for *Challenger*, is rolled out at Air Force Plant 42

KEY: *Mission* or *Spacecraft name*, (Cosmonaut or Astronaut)

FOREWORD

On the launch pad of the arid steppes of Kazakhstan, a huge rocket was waiting to propel Yuri Gagarin, the first man in space, into the vast universe. Before departure, Gagarin greeted an excellent group of colleagues. "Dear friends, you who are close to me, and you whom I do not know. Fellow Russians, and people of all countries and continents: In a few moments a powerful space vehicle will carry me into the distant realm of space. What can I tell you in these last minutes before the launch? My whole life now appears to me as one beautiful moment..." Gagarin's words on April 12, 1961, mark the beginning of many more celebrated moments and great achievements of the Soviet space program.

Just four short years later, on March 18, 1965, I was fortunate enough to be the first man to walk in space when I floated outside my *Voskhod 2* spacecraft for ten incredible minutes. The beauty of my experience cannot be described.

The American and Soviet space programs came together in July of 1975 for the notable *Soyuz-Apollo* mission. Valery Kubasov, a flight engineer, and I spent two days with three American astronauts, Deke Slayton, Tom Stafford and Vance Brand, in our joined *Soyuz-Apollo* spacecrafts. We proved that, politics aside, people from two great countries can achieve great things working together in space.

The citizens of our two countries are happy about the end of the Cold War. We have high hopes that *glasnost* and *perestroika* will continue to bring positive changes to our nations. Only in recent years have these developments allowed for an exhibit as remarkable as *Soviet Space* to come to Fort Worth, Texas. We are proud to represent our good feelings and wishes for future cooperation in all areas between the United States and the Soviet Union through this Exhibit. The history of *Soviet Space* and the technological developments are close to my heart. I invite these same thoughts and dreams to become close to the hearts of Texans as well.

The Russian heart is a vast one, as vast as the universe itself. I think Texans must have a great heart as well. We are grateful to the Fort Worth Museum of Science and History for providing the means to display our Exhibit. Our hearts, together with the heart of the universe, welcome all with the desire to launch dreams into the realms beyond our planet Earth.

Alexei Leonov
Deputy Head of Flight and Space Training
Yuri Gagarin Cosmonaut Training Center

PROLOGUE

On the gossamer threads of a dream, we have traveled—in the span of a single human lifetime—from featherweight wood-and-fabric bi-planes to multi-thousand-ton spacecraft. We have, in the most literal turn of the phrase, "burst the bonds of Earth." We have begun to explore the infinite universe.

Why, one might ask, should we embark on such a risky venture at this point in human history? In a world torn with strife, hunger, disease, diminishing natural resources, and seemingly endless discord, why should we expend still more of our precious physical, intellectual, and economic resources on a technology that gives every impression of benefiting only a privileged few?

Why?

Precisely because exploration of the cosmos has the potential of benefiting us all. It offers in one rather complex but clearly defined package, the solution to virtually every economic, social, technological, and natural resource difficulty presently facing the Earth and its nearly six billion human inhabitants. It is the key to a future of unlimited energy, infinite natural and food resources, and unimagined territorial expansion options. There is in the cosmos so much of everything, not even man's innate penchant for greed can stand in the way of its exploration.

Like Columbus, we presently sit on the threshold of a new world—not yet knowing exactly where it is or how to get there, but seemingly willing to take the chance. This exhibit, *Soviet Space*, represents that willingness—and concomitantly, the first tentative steps toward cosmic exploration. It represents, with considerable clarity, both the dreams and the realities of a world that might be . . . a world of cooperation rather than confrontation, hope rather than despair, and feast rather than famine. In a world with skewed social priorities and ever increasing chaos, exploration of the cosmos is our greatest hope; the alternative is simply too terrifying to contemplate.

Jay Miller

« Человечество не останется навсегда прикованным к Земле. В погоне за светом и пространством оно, сначала робко, исследует пределы атмосферы, а затем овладеет всей солнечной системой. »

Константин Циолковский

"Humanity will not remain on Earth forever; it will chase after light and space, then timidly move out beyond the atmosphere, and ultimately conquer the entire solar system."

Konstantin Tsiolkovsky

EARLY SOVIET ROCKETRY

Soviet intellectual giant Konstantin Eduardovich Tsiolkovsky looms over the history of space travel not only as its earliest and most ardent proponent, but its most accurate prophet as well. Born on September 17, 1857, in the village of Izhevskoye, he grew up impoverished, deaf (as a result of scarlet fever), and almost totally self-educated. He later made his living as a provincial school teacher in the small town of Kaluga (approximately 125 miles [201 km] southwest of Moscow) and eventually developed a strong affinity for the rarefied atmosphere of science and technology.

In particular, Tsiolkovsky became enamored of rocketry and space travel. Because of circumstance, he centered his early studies not on hardware research and an interchange of ideas with colleagues, but rather on cosmic dreams and the meager references available from local libraries and miscellaneous other sources.

Tsiolkovsky's dreams became the embodiment of other dreamers' imaginations. Most notably, the science fiction book *De la terre a' la lune (From The Earth To The Moon)* written by the great French author Jules Verne, caught his undivided attention. Much later in life, he would note that Verne's " . . . fantastic novels . . ." served as his " . . . first seeds of thought . . ." about space flight; " . . . they set my brain working in a definite direction. I mused; musing led me to more serious mental activity."

Tsiolkovsky, though eventually to achieve a level of great prominence in the small Soviet space community of the 1920s and 1930s, as a result of the writings of Nikolai Rynin and Yakov Perelman, remained a relatively obscure figure to Western rocketry proponents until the translation and publication of select Tsiolkovsky writings in Western journals during the late 1940s. In the interim, U.S. and German space giants Robert H. Goddard and Hermann Oberth achieved their own levels of prominence—though in space enthusiast communities that remained all but invisible until the advent of World War II.

(right) *The father of Soviet space flight, Konstantin Eduardovich Tsiolkovsky, was born in 1857 and died in 1935. A man of remarkable intellect and foresight, his writings and theories affected the lives of many space pioneers. Today he is held in great esteem by space theoreticians and historians around the world.*

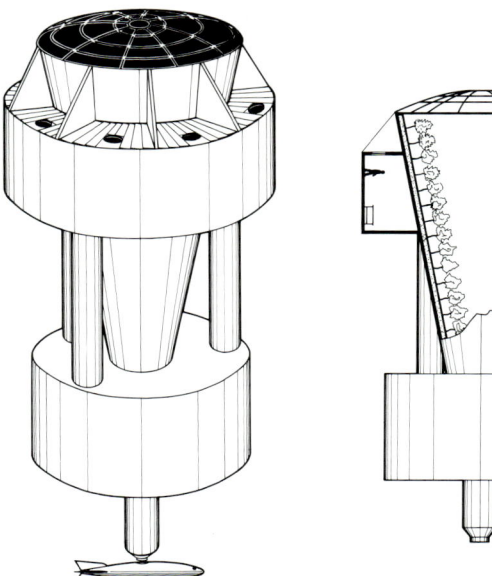

Though Tsiolkovsky's work initially had little impact on Western rocket and space research (only 7,500 Tsiolkovsky books ever were printed—and precious few of those ever reached the West), the fact that he began accurately predicting space travel and its many applications as early as 1883 placed him at the forefront of the small cadre of truly forward thinking space prophets. Perhaps more importantly, his precedent setting 1903 article "The Investigation of Universal Space By Means of Reactive Devices" set forth the fundamental physical laws of reactive motion in space (a rocket accelerates in powered flight because it is constantly losing weight as the propellant is burned). He also accurately predicted that liquid oxygen and liquid hydrogen would become the preeminent rocket propellants of choice.

The reactive motion law and the propellant prediction were unerringly accurate deductions that elevated Tsiolkovsky from a mere science fiction writer to a giant among his astronautical peers. The publication of those insights proved beyond doubt that he not only was working at a previously unexplored level of space dynamics, but also at the leading edge of space technology. Because of a general lack of official appreciation for their importance, Tsiolkovsky's conclusions were considered public domain and not proprietary to Czarist Russia; other scientists could exploit them at will. Unfortunately, the great scientist's poverty and the limited communications networks of his day curtailed the rapid international acknowledgment of his genius. At the time of Tsiolkovsky's death in 1935, only the most learned of Western space proponents were aware of his monumental contributions.

Because of its small following, the international space community of the 1920s and 1930s was highly dependent upon publicity in its haphazard attempts to gain support and acquire research financing. Space flight was still an avant-garde science; resources for theoretical and/or hardware research in the post-revolutionary Soviet Union were extremely difficult to acquire.

(above) *Tsiolkovsky, though primarily a writer and theoretician, did experiment with miscellaneous bits and pieces of actual space hardware. Many of his paper design studies, including the somewhat more pedestrian rigid dirigible on the right in this photo, were built in model form.*

(middle) *The precedent-setting GIRD-X liquid-fuel rocket and several members of the Moscow GIRD rocket society including on the far left, Sergei Korolev. It is thought that other identifiable members probably are Leonid Dushkin (with glasses); Leonid Korneyev (to right of rocket, standing); Alexander Polyarny (crouched in middle row, far right); and L. Kolbasicha (the sole woman attendee).*

(left) *An original space station configuration as envisioned by Tsiolkovsky. This early study incorporated a closed ecological system, a garden, a laboratory, passages to permit access to storage and living areas, a docking mechanism, and even a supply-type space shuttle.*

EXHIBIT ARTIFACT: A small-scale model of one of Tsiolkovsky's manned space rocket designs. The upper-three levels are designed to accommodate the crew members and their research equipment. The bottom half is utilized to serve as the propellant tank. The liquid-hydrogen and liquid-oxygen powered rocket engine is centrally mounted and exhausts out the vehicle's base. Small "impingement vanes" project into the exhaust flow and when moved upon pilot command, direct the rocket's course.

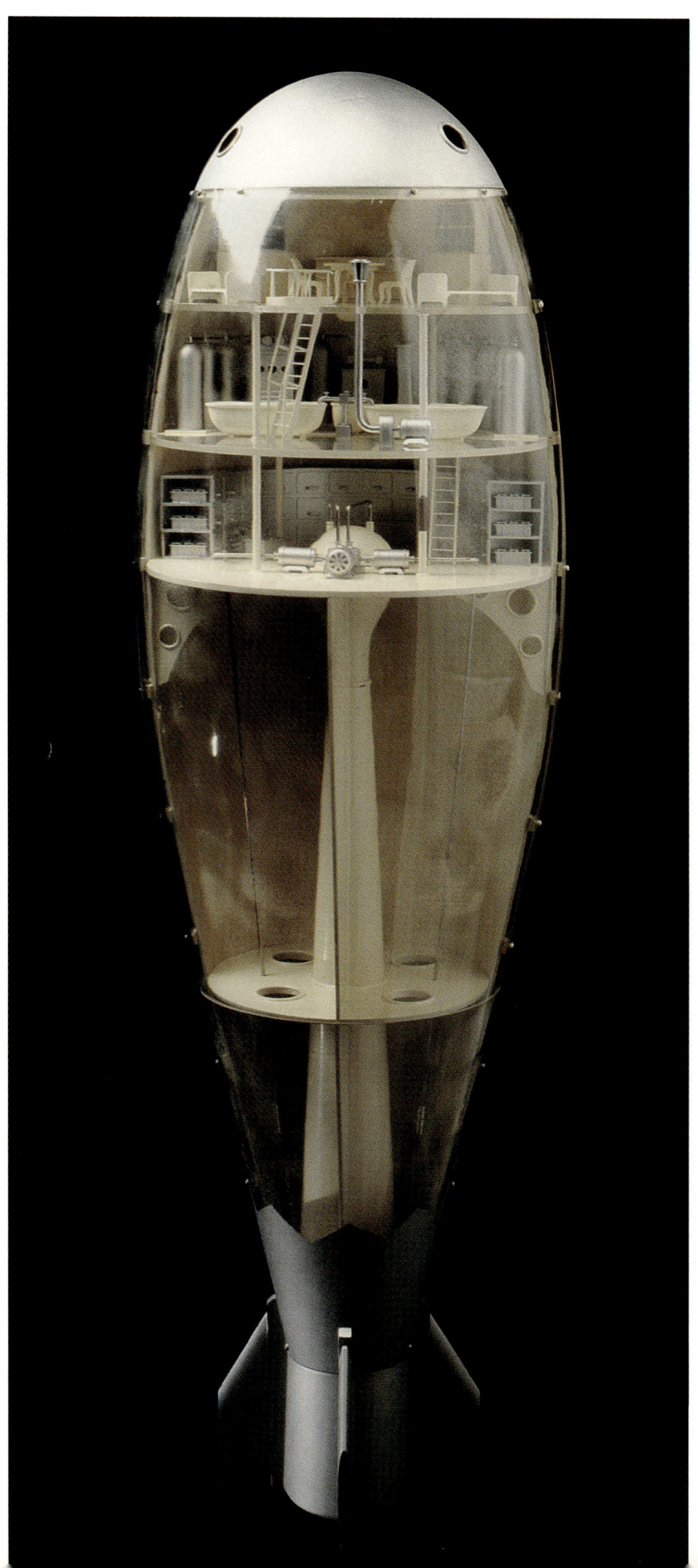

EXHIBIT ARTIFACT: *The experimental R-06 (top) and GIRD-X (bottom) rockets. Nine R-06s were built and launched during 1937 and 1938 by the Osoaviakhim/KB-7. GIRD-X was the first successful Soviet liquid-propellant rocket. It was launched on November 25, 1933, and reached an altitude of over 246 ft [75 m].*

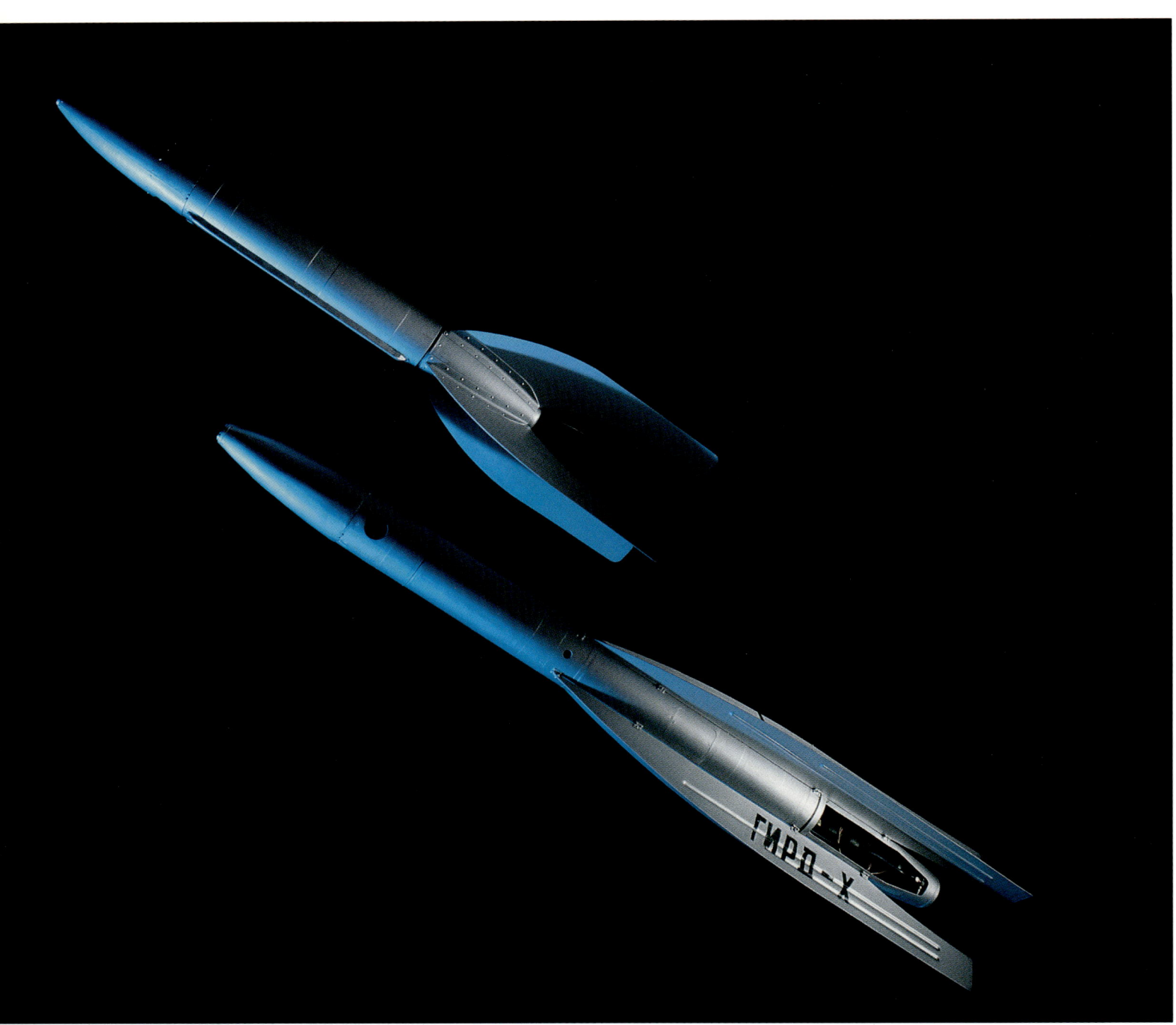

(top) Design Office 7 rocket R-06
Full-scale replica
Launch weight: 22 lb [10 kg]

(bottom) GIRD-X rocket
Full-scale replica
Launch weight: 65 lb [29.5 kg]

EXHIBIT ARTIFACT: A 1:10 scale model of the GIRD-X starting stand. The rocket was placed inside of the central rail assembly and launched vertically following rocket engine ignition.

Frederick Arturovich Tsander was a gifted speaker, a tireless proponent of space flight, and an ardent disciple of Tsiolkovsky and his space travel theories. Whereas Tsiolkovsky was inarguably cerebral and more adept with a pen than a speaker's stand, the decidedly extroverted Tsander was markedly charismatic and personable. As Tsander gained notoriety in the several Soviet space societies, his ability to stimulate interest in what then was considered an esoteric subject was put to considerable test. Giving speeches about space and space flight brought him considerable acclaim, and on at least one occasion, his audience included none other than Marshall and General Secretary of the Soviet Union's Communist Party Joseph Stalin.

Tsander's knowledge, coupled with the respect he was given by astronautical enthusiasts throughout the Soviet Union, inadvertently led to his exposing the Eastern space community to the Western. His knowledge of Goddard and Oberth and their significant achievements in the U.S. and Germany, served as a considerable revelation to many Soviet space proponents.

It was, in fact, a speech given by Tsander on January 20, 1924, that birthed what would become an extensive network of Soviet "rocket societies." Created to further man's knowledge and interest in rocketry and space exploration, the first of these, referred to as the *Obschestvo po izutcheniyu mezhplanetnykh puteshestvii* ("Society for the Study of Interplanetary Travel"), by April had been absorbed formally into the Military Science Division of the N.E. Zhukovsky Air Force Academy in Moscow. This military association proved short lived, however, and within one month, the society had reorganized privately as the *Obschestvo po izutcheniyu mezhplanetnykh soobschenii* ("Society for the Study of Interplanetary Communication" or OIMS).

As word of the society spread, many noteworthy Soviet space pioneers became active in its proceedings. Tsander was joined, at least in spirit, by the great Tsiolkovsky (whose health and distant residence prevented his attending meetings); by the argumentative and knowledgeable university professor and space proponent Lapirov-Skobolo; and by the prominent Professor Vladimir P. Vetchinkin—whose name would become deeply enmeshed with the history of Soviet astronautics and associated academics during the many formative years that lay ahead.

Debate within the Society over misreported news that U.S. scientist Robert Goddard had landed a missile on the Moon on August 5, 1924, eventually led to the organization's temporary dissolution. Moscow newspapers, like those in the U.S., had misrepresented the thrust and intellectual importance of the Goddard interview—which incorrectly implied Goddard had successfully built and flown rockets that were designed to carry men to the Moon. Soviet readers chastised Society members and accused them of being cranks. Disgraced and ridiculed, they disbanded.

The original Society was followed by others, including one assembled in Kiev in the Ukraine under the aegis of academician D. A. Grave. This latter group laid claim in their hometown, on June 19, 1925, to being the first to hold a major space flight exhibition. Almost two years later, a similar though considerably larger exhibit, referred to as the "First World Exhibition of Interplanetary Machines and Mechanisms", was assembled in Moscow. This display opened to the public during April of 1927 and included special displays acknowledging the contributions of authors Jules Verne and H. G. Wells, as well as space pioneers, Tsiolkovsky, Goddard, and Oberth. Interplanetary rockets (including one proposal to be powered by nuclear propulsion units), a spacesuit, and radio communications were some of the advanced technologies displayed.

With the closing of Moscow's "First World Exhibition" in June of 1927, a number of informally organized enthusiasts in Leningrad and other Soviet cities began a more serious thrust in the direction of proper space societies. By late 1928, Nikolai Alexeyvich Rynin, author of a highly acclaimed encyclopedia of astronautics entitled *Interplanetary Communications*, had succeeded in assembling a group under the sponsorship of the "Leningrad Institute of Communication Engineers." Somewhat earlier, a similar group had evolved from Vladimir Artemev's powder rocket research team to become, by 1929, the militarily-oriented Gas Dynamics Laboratory (GDL) in Moscow.

The GDL, headed by chemical engineer Nikolai Ivanovich Tikhomirov and under the general supervision of Mikhail Tukhachevsky, also eventually found itself in Leningrad. With input from the brilliant mind of the newly arrived team member Valentin Petrovich Glushko, the small GDL staff soon was exploring not only powder-fueled anti-aircraft and anti-tank rockets, but also rudimentary liquid-propellant rockets.

Glushko, in 1924 at age 18, had written articles describing space flight, space stations, and artificial satellites. By the time of his acceptance into the GDL at the age of 21, Glushko had become a recognized authority on the subject of astronautics and rocket propulsion.

During 1931, two new Soviet rocket societies surfaced; Moscow's *Gruppa po izucheniyu reaktivnogo dvizhenia* ("Group for the Study of Reactive Motion" popularly referred to as the MosGIRD) and its Leningrad sister organization, the LenGIRD. Falling under the jurisdiction of the *Osoaviakhim* ("Society for Assisting Defense and Aviation and Chemical Construction in the USSR"), MosGIRD and LenGIRD proved important nookeries for such future Soviet space greats as Tsander and a neophyte engineer by the name of Sergei Pavlovich Korolev.

With MosGIRD gaining the stature of a formal order of establishment during July of 1932, the group turned to its newly elected chief, Korolev for guidance and direction. Korolev, along with other members of the society, already had full-time employment in the government aircraft industry. He agreed, however, that MosGIRD should not only be supported, but utilized as a think tank and hardware development facility for astronautical projects.

By this time, Tsander, also a MosGIRD member, had begun development of early liquid-fuel rocket engines and had created his novel OR-1 test unit from a "borrowed blowtorch." Paralleling his effort, though done under the umbrella of the GDL and its penchant for military secrecy, was Glushko's *Optynyy raketnyy motor* number 1 ("Experimental Rocket Engine #1" or ORM-1). This later would be acknowledged as the Soviet Union's first viable liquid-fuel rocket engine.

The ORM-1 would birth a large family of successful GDL-sponsored liquid-fueled rocket engines. These would continue in development under a great veil of secrecy even while World War II raged with great fury across the land. Their practical applications would prove few and far between, however, and their contributions to the war effort would be virtually non-existent.

Tsander and Korolev now forged ahead with a gutsy and historic plan of their own—to build, under the auspices of MosGIRD, the Korolev-designed RP-1 aircraft and power it with Tsander's OR-2 "experimental rocket engine." Completion and flight test were to take place by January of 1932.

By now, the GIRD had grown to considerable proportions and consisted of an eclectic membership with many differing ideas as to what the organization should represent and what its objectives should be. As a result of a rapidly growing awareness in the military of the potential of rocket propulsion systems and missile technology, the GIRD was somewhat forcefully merged with the GDL during 1932, and consequently, fell under military jurisdiction.

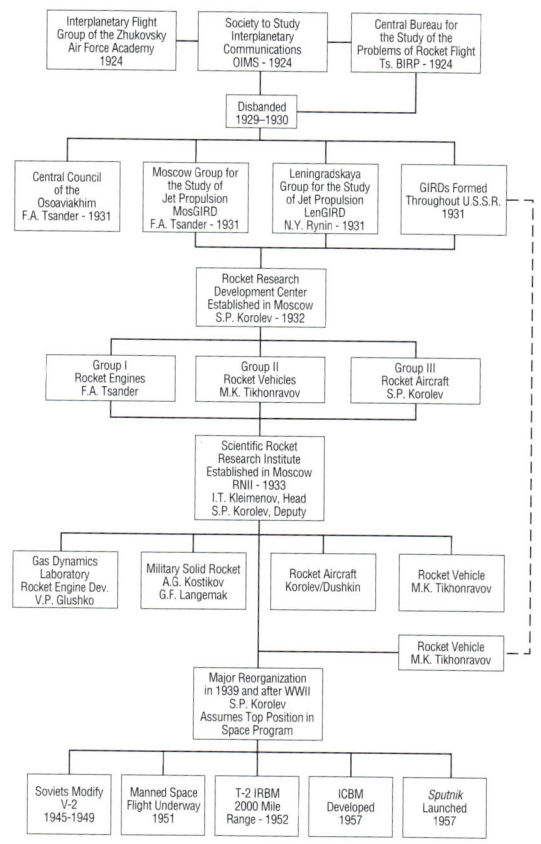

(above) *Soviet rocket pioneer Alexander Polyarny during 1935 holding what appears to be the R-06 rocket. In the background are various members of the Stratospheric Committee of the U.S.S.R. Osoaviakhim Central Council. This was just one of several fledgling Soviet space enthusiast organizations that later led to the formation of the Soviet space program.*

(right) *Evolutionary chart provides a family tree of all known Soviet rocket societies and their miscellaneous spin-offs, including the birth of the Soviet ICBM program and the 1957 launch of Sputnik 1.*

«Кто, устремляя в ясную осеннюю ночь свои свои взоры к небу, при виде сверкающих на нем звезд не думал о том, что там, на далеких планетах, может быть, живут подобные нам разумные существа, опередившие нас в культуре на многие тысячи лет. Какие несметные культурные ценности могли бы быть доставлены на земной шар, земной науке, если бы удалось туда перелететь человеку, и какую минимальную затрату надо произвести на такое великое дело в сравнении с тем, что бесполезно тратится человеком.»

Фридрих Цандер

"Who, while lifting his gaze skyward to the sight of stars sparkling in a clear autumn night, has never thought that there live on far-off planets rational beings similar to ourselves who are far advanced of us by many thousands of years. What countless cultural treasures could be brought back to Earth, to earthly science were man able to fly to them and return and what a minimal cost there would be for such a great feat in comparison with what man is accustomed to spending in fruitless effort."

Frederick Tsander

By August of 1932, the GDL/GIRD merger had been approved and consummated by Tsander, Korolev, and other leading society members. The military perspective now began to infiltrate the general philosophical thrust of the group and eventually the dream of space flight began to wither. Though wages, accommodations, and general funding improved dramatically, it was apparent that the lofty goals ascribed to by the original GIRD members were rapidly giving way to the more pragmatic requirements of war. Rather than aspire to trips to the Moon and the stars, thoughts turned toward development of effective high-altitude anti-aircraft and surface-to-surface missiles.

In the meantime, though the Tsander/Korolev RP-1 aircraft was completed and flown on several occasions as an unpowered glider, it was never to become airborne powered by Tsander's rocket engine. Tsander's untimely death on March 28, 1933, the changing philosophical climate of the revamped Society, and the wear and tear that the airframe had suffered during initial testing all combined to expedite a decision to terminate the project.

Though work on the RP-1 had ended, parallel Russian rocket programs were just beginning to reach fruition in the form of rocket engines and applicable missile bodies. Among these was the successful construction and launch of the Soviet Union's first hybrid-fuel (liquid oxygen and jellied gasoline) rocket, the GIRD-09 on August 17, 1933. This was followed, on November 25, by the posthumous launch of Tsander's liquid fuel (liquid oxygen and alcohol) GIRD-X.

The previous month, on October 31, 1933 a state resolution creating the *Reaktivni nauchno-issledovatel'kii institut* ("Scientific Research Institute of Jet Propulsion"), more commonly referred to as the RNII, became what the Soviet bureaucracy referred to as the first formal rocket research establishment in the world. Assigned as head of this new operation was Ivan Terentevich Kleimenov and perhaps more importantly, his deputy, the 29-year-old, newly decreed Major General Sergei Pavlovich Korolev.

(left) *The GIRD society's R-07 rocket was designed during 1933, and launched on November 17, 1934. This vehicle represented a most unusual configuration. The rocket engine was mounted above the normal center of gravity and the propellant tanks (two containing alcohol and two containing oxygen) were housed inside the extended stabilizing fins.*

Integral with the R-7 program and critical to its success was the parallel development of launch and support facilities some 1,500 miles [2,407 km] to the southeast of Moscow near the railway town of Tyuratam, later to become known to the world as the Baikonur Cosmodrome. Baikonur would remain the primary launch site for all rockets carrying manned, lunar, planetary, geosynchronous, tactical ocean surveillance, high-altitude navigation, anti-satellite, and some photographic reconnaissance space vehicles.

Two other cosmodromes achieved certain levels of notoriety, these being located some 250 miles [402 km] north of Moscow near the town of Plesetsk and some 500 miles [805 km] south and slightly east of Moscow near the town of Kapustin Yar. Their integration into the launch program, however, was primarily military and strictly hardware-oriented. The Kapustin Yar facility (in the republic of Kazakhstan) had actually been the site of the first Soviet V-2 ballistic missile test on October 30, 1947, and thus was the oldest active cosmodrome. Plesetsk, on the other hand, was the newest, having been integrated into the cosmodrome complex during the early 1960s.

With Baikonur, Plesetsk, Kapustin Yar and numerous other research centers actively involved in rocket propulsion and missile design by 1953, development of bigger and better launch vehicles moved ahead in the Soviet Union at an extremely rapid pace. Not surprisingly, the West was oblivious to this activity; not one word of the massive technological investment encompassed by the Soviet missile program had been leaked to the Western press, and the intelligence community was equally hard pressed for facts. ●

(right) *The Soviet Union was overflown illegally some 30 times by U.S. Central Intelligence Agency-operated U-2 high-altitude reconnaissance aircraft between 1956 and 1960. This image, taken during the course of one of those flights, depicts an R-7 ICBM launch complex at Tyuratam. Boosters were brought to such pads by rail (visible at bottom) and were erected following their arrival. A massive flame pit is visible in the image's upper half.*

«Выход за пределы Земли, который открыла космонавтика, затрагивает каждого из нас и вселяет надежду, что со временем, быть может, мы не только научимся лучше понимать мир, окружающий нас, но и самих себя.»

Константин Феоктистов

"The ability to leave the Earth, which space flight has given us, moves us all and instills the hope that, perhaps, given enough time, we will not only learn to better understand the world around us, but also ourselves."

Konstantin Feoktistov

The venerable Korolev's importance to the Soviet space program can not be over-stated. From relatively humble beginnings as a proponent of the space societies and the designer of rudimentary rocket engines and rocket-propelled gliders, to the birth of the Soviet intercontinental ballistic missile (ICBM), space station, and Moon rocket programs, Korolev's influence, control, and great genius were without peer. Until his death in January of 1966, he was instrumental in the design and engineering of virtually every important Soviet space vehicle of his day. As "Chief Designer of Carrier Rockets and Spacecraft" from the early 1950s until his death, he now is acknowledged to have been the single most important management and engineering personality in the post-WWII Soviet space program.

Peripheral, but related to the development of Korolev's ICBM, was the associated development of Soviet nuclear weapons. The successful detonation of the first such Soviet device, during August of 1949, set in motion the development of a warhead that was missile transportable. Each of these efforts became, in effect, driven by the other, as the warhead designers now began to attempt to reduce the size of the nuclear bomb to a device that could be accommodated by the launch vehicle, and the launch vehicle designers now began to attempt to increase the size of their missile to accommodate the device. Inevitably, the two objectives were achieved, and by the date of Korolev's winning ICBM design submission during 1954, were beginning to coalesce into the Soviet Union's most successful and longest lived space booster, the R-7 (eventually to become known in the West as the SS-6 *Sapwood* ICBM and in its space capsule launcher form, as the SL-1).

When first tested during 1956, the R-7 represented the synthesis of virtually everything known in the Soviet Union about space and rocketry. Korolev's dream, built on the shoulders of his direct and personal relationships with Tsiolkovsky, Tsander, the GIRDs, the GDL, the RNII, and the various German space scientists, had finally reached fruition in one huge, but immensely practical rocket.

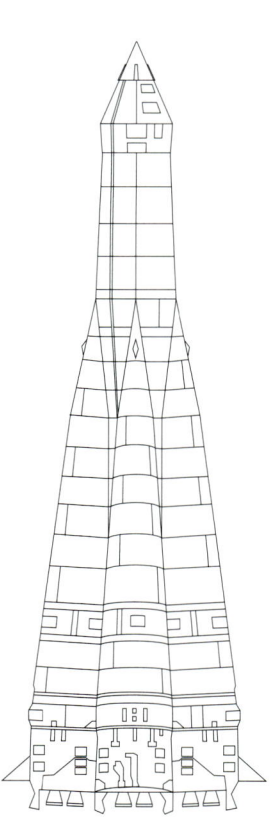

The beauty of the R-7 was its relative simplicity and its basically utilitarian nature. Modular both in concept and capability, it could be utilized not only for lobbing thousand-pound nuclear warheads over great distances, but also for lofting cumbersome scientific payloads into Earth orbit. Based broadly on the technology that had been gleaned from the old German V-2 rocket, the R-7 eventually would prove to be safe, inexpensive, and most importantly, very reliable.

(above right) *The R-7 was the Soviet Union's first intercontinental ballistic missile.*
(above) *Drawing of the R-7, configured as a space launch vehicle.*

In 1947, with the specter of this threat as its justification, the *Pravitel'stvennaya komissiya po raketam dalnego deistviya* (P.K.R.D.D.—State Commission for the Study of the Problems of Long-Range Rockets) was formed and tasked with the study of large rockets and how they might be utilized to destroy targets at extremely long ranges. The objective was to determine the feasibility of building what eventually would be referred to as intercontinental ballistic missiles, or ICBMs—and using them to loft warheads over intercontinental ranges.

By early 1948, Soviet-built V-2s were being launched with considerable regularity and utilized for a variety of atmospheric research and hardware development programs. The latter were under the control of Yuri A. Pobedonostev—a former TsAGI ("Institute for Aerodynamic and Hydrodynamic Research") and RNII engineer who had been tasked after the war with exploration of V-2 technology. Consequent to this, an improved, longer-range version of the V-2 was developed under Pobedonostev's direction and later referred to as the *Pobeda* ("Victory") or R-14. This missile, alongside conventional V-2s, during 1950 and 1951, served as the basis for the formation of the first Soviet combat rocket division.

Concurrent with the work being conducted on intercontinental range missiles during this period, the Soviets also developed a number of highly reliable atmospheric and geophysical research rockets, including the 1950s vintage MR-1, or *Meteo*. This missile could explore the atmosphere to an altitude of 60 miles [97 km] while carrying scientific payloads weighing up to 176 lb [80 kg]. Of advanced design, it utilized a solid-fuel booster and a liquid-fuel sustainer stage to propel its research package into near-space. The package then was ejected and returned to Earth by parachute. Later, similar but more advanced Soviet research rockets such as the V-2-A (an advanced version of the original German V-2) reached altitudes of 132 miles [212 km] while carrying atmospheric research payloads weighing almost 5,000 lb [2,268 kg].

By 1954, the successes that had been enjoyed by the Soviet space community had led to considerable Kremlin support. Coupled with what by now had become an adversarial relationship with the West, the incentive to forge ahead with plans for the development of a truly intercontinental range ballistic missile—capable of delivering accurately a warhead onto a distant target—was sufficient to merit the acceptance of a related design submission from a team headed by purge-survivor Sergei Korolev.

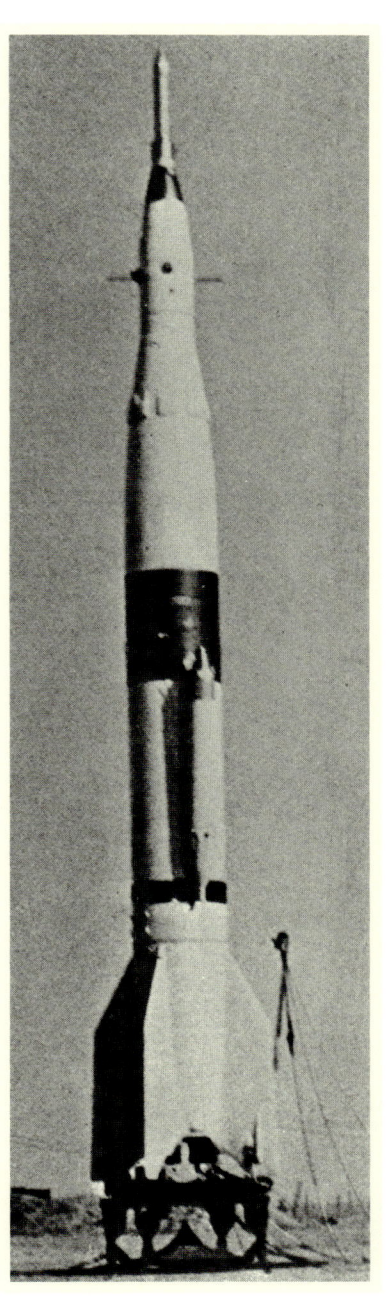

(above) *This Soviet geophysics rocket strongly resembled the German V-2, from which it was derived.*

this did not go unnoticed by allied military observers. The V-2 barrage continued until March 27, 1945, when allied encroachment curtailed further launches.

By the end of World War II, the V-2's successes had forever changed the face of war. Soviet troops had captured Peenemunde and a large quantity of hardware; U.S. troops had overrun the Harz Mountains V-2 central production plant; and both allied combatants had captured a considerable number of German rocket scientists. These acquisitions would serve the U.S. and the Soviet Union well, as they would become the seeds from which exploration of the cosmos would grow.

Though the occupation of Peenemunde proved a major windfall for the Soviet rocket teams, it was not without serious shortcomings. For one, the U.S. and its Western allies had acquired the majority of the more accomplished German rocket scientists (including Walter Dornberger and Wernher von Braun), and for another, much of the fixed hardware at the German test facility had been damaged or destroyed by allied bombing attacks.

Stalin now ordered Peenemunde's launch pads and testing facilities rebuilt, and within a year, German Helmut Gröttrup, one of the few upper-echelon scientists captured by the Soviets, had managed to restart V-2 production and recreate most of the test site's research laboratories. This effort proved in vain, however, for within weeks of its completion, all Peenemunde personnel, consisting of many hundreds of engineers, scientists, workers, and their families, were arbitrarily and inexplicably ordered to pack and prepare for relocation.

Shortly afterward, the Germans were moved to a site near Moscow and effectively placed in isolation. Their involvement in Soviet rocket programs was limited to an "as needed" basis, and their "hands-on" contributions rapidly were reduced to zero. Soviet rocket technology, by that time, had superseded German, and coupled with a prevailing anti-German sentiment, later led to a decision to return the Peenemunde personnel to their homeland. The last group left for Germany during November of 1953.

By 1947, post-war Soviet rocket research activity had returned to a modest level of productivity and Soviet-built V-2s were being manufactured at rates considerably higher than those seen in Germany at the end of the war. In the interim, the political and philosophical climate had begun to change in the Kremlin, and the general consensus of opinion among those in its hierarchy was that the western powers were evolving into the only major threat to Russian sovereignty.

« С берега Вселенной, которым стала священная земля нашей Родины, не раз уйдут в неизведанные дали советские корабли, поднимаемые мощными ракетами-носителями. Их полет и возвращение будут великим праздником советского народа, всего передового человечества — победой разума и прогресса. »

Сергей Королев

"From the edge of the universe, which is now located on the hallowed soil of our Motherland, Soviet spaceships, hoisted aloft by powerful booster rockets, will depart for uncharted destinations. Their departure and return will be a great holiday for the Soviet people and for all forward-looking humanity, a victory for reason and progress."

Sergei Korolev

At the beginning of World War II, the state of international rocketry remained one of experimentation and military exploitation. In the U.S., Robert Goddard's work had been only tacitly acknowledged and little support had been forthcoming from government agencies; rocket societies in several European countries including England had met with only modest success and limited membership; and only Germany had managed to move from novelty to significant government interest.

In some respects, the birth of German rocketry effectively mirrored that of the post-revolutionary Soviet Union. Space societies and organizations came and went, government and private support proved tenuous, and it was only after the discovery that rocket propulsion could be applied to weapons of war that serious interest finally was shown.

Though historically starting later than its Soviet counterpart, German interest in rocketry gathered steam quickly. By 1937, government support was strong enough to provide a well-equipped and adequately manned test facility near the small German town of Wolgast. Known formally as "Army Experimental Station Peenemunde," it eventually became known simply as Peenemunde; over the next eight years it would become the foremost rocket and missile research center in the world.

German military successes resulting from the use of conventional combat equipment and the infamous "blitzkreig" attack strategy in Poland and France at the beginning of World War II led to a rapid and major de-emphasis of the "high-tech" rockets and missiles being developed at Peenemunde after 1940. On October 3, 1942, however, an experimental liquid-fuel V-2 (*Vergeltungswaffe-zwei* or "Vengeance Weapon 2") rocket was launched successfully, eventually hitting a predetermined surface contact point some 120 miles [193 km] from its Peenemunde lift-off pad. Word of this accomplishment, coupled with an adverse shift in war fortunes then being assessed by Hitler and his deputies, led to a re-evaluation of Peenemunde rockets and a rapid decision to place the V-2 into production.

On September 6, 1944, the first of some 4,300 V-2s was launched by the Nazis against England and surrounding European countries. At least 1,500 eventually headed in the direction of British soil, and an additional 2,100 went skyward seeking targets across Europe. The 3,000 mph [4,828 km/h] missiles had a devastating psychological and physical impact on the British and European populations, and the importance of

It is appropriate, at this point, to recount the most memorable Soviet pre-war space pioneers, space society members, GIRD members, and GDL members and their respective contributions; many of these men later would play prominent, if not critical roles in post-war Soviet rocketry and space travel:

Artemev, V.A.
solid-fuel rocket propulsion systems specialist

Chaplygin, S.A.
gas dynamics and rocket propulsion specialist

Dushkin, L.S.
rocket-propelled aircraft and rocket specialist

Fedorov, A.P.
early rocketry and space travel theorist

Glushko, V.P.
liquid-fuel rocket engine specialist

Grave, D.A.
academician and Kiev space society founder

Gryaznov, V.
rocket engine development specialist

Isayev, A.V.
rocket engine development specialist

Kibalchich, N.
formulated plans for gunpowder rockets

Kondratyuk, Y.V.
early rocketry and space travel theorist

Korneyev, L.K.
propellant combustion technology specialist

Korolev, S.P.
aircraft, booster, and spacecraft design specialist

Kostikov, A.G.
solid-fuel military rockets and related propulsion

Langemak, G.D.
solid-fuel rocket propulsion systems specialist

Lapirov-Skobolo, M.I.
professor and early space flight proponent

Merkulov, I.A.
ramjet engine design specialist

Meshchersky, I.
specialist in theoretical mechanics and rocketry

Moshkin, E.K.
rocket engine design specialist

Perelman, Y.I.
space author

Petropavlovsky, B.S.
solid-fuel rocket propulsion systems specialist

Pobedonostev, Y.A.
rocket booster design specialist

Polyarny, A.I.
propellant combustion technology specialist

Razumov, V.V.
rocket booster design specialist

Rynin, N.A.
space author

Salikov, A.V.
rocket engine development specialist

Shtern, A.N.
rocket engine design specialist

Stechkin, B.
jet and ramjet engine specialist

Tikhomirov, N.I.
rocket designer

Tikhonravov, M.K.
rocket engine and vehicle design specialist

Tsander, F.A.
rocket engine/spacecraft pioneer and proponent

Tsiolkovsky, K.E.
father of Soviet space flight

Vetchinkin, V.P.
professor and early space flight proponent

Zhukovsky, N.E.
father of Soviet aviation

With its goals of atmospheric and stratospheric exploration set at a more practical level, the RNII began development of numerous sounding rockets (designed to carry sampling payloads into the upper atmosphere), "winged rockets" (a Soviet euphemism for the conventional tailed rocket and missile configurations that today are considered commonplace), and ramjet propulsion systems. Concurrently, during the summer of 1935, another Soviet rocket organization, Design Bureau No.7 (KB-7), was created. With military connections, it apparently was formed with the intent of designing and building rocket-powered weapons for combat.

Outside of the RNII and KB-7, many independent rocket and missile projects were born during this seemingly fertile period. Some of these, like the 214 lb [97 kg] *Aviavnito* sounding rocket developed by the *Osoaviakim* "All-Union Aviation, Scientific, Engineering and Technical Society," were modestly successful. The vast majority were aborted prior to completion, however, or failed in the hardware stage.

As far back as 1934, Joseph Stalin had embarked on what now is referred to as "The Great Purge." In a shameful attempt to cleanse the Soviet bureaucracy and general populace the Soviet dictator arbitrarily deemed disloyal to the communist cause, a semi-systematic slaughtering of government personnel and civilians was undertaken that eventually resulted in the deaths of millions. The various space societies and rocket organizations, both civil and military, were not immune to this "thinning of the ranks," and among those court-martialled and shot were thousands of engineers, scientists, and other highly trained and skilled aerospace professionals.

Prominent space and rocketry proponents who lost their lives during this inexplicable period included the "grand patron of Soviet rocketry," Mikhail Tukhachevsky; the chief of the RNII, I. T. Kleimenov; the deputy chief of the RNII, Georgi Erickhovich Langemak; and the chief of the GDL, Nikolai Yakovlevich Il'yin. These were dark days in the Soviet Union, and their affect on Russia, Soviet society, and the Soviet bureaucracy would be everlasting.

By 1938, fear and paranoia raged unabated throughout the Soviet Union. Making matters worse was the looming specter of world war. A pall of depression had fallen over all of Russia. The space and rocket societies and related bureaucratic organizations suffered enormously and, in the case of the GIRDs and RNII, all but disappeared. Voluntary membership—at approximately one-thousand, never very large—virtually evaporated as a result of "guilt by association" with Tukhachevsky—who not only had become supervisor of the GDL, but also the head of the Revolutionary Military Council's Department of Armaments, Marshall of the Soviet Union, and Deputy Minister of Defense.

By 1939, the only viable remnant of the original rocket societies was the RNII's GDL sub-unit. This innately pro-military operation now was integrated with the Moscow Aviation Engine Plant. During 1941, under the direction of Glushko and his deputy, Korolev, it became the GDL-OKB (Experimental Design Bureau). As such, it concentrated on the design and development of liquid-fuel rocket engines, rocket-assisted-takeoff (RATO) units for conventional aircraft, and advanced rocket propulsion systems for air-to-surface and surface-to-surface missiles.

At the beginning of Soviet involvement in World War II—on September 1, 1939—among the several GDL-OKB surface-to-surface missile projects underway was the *Katyusha* ("Little Katy") rocket being developed by A. G. Kostikov. This simple, solid-fuel weapon, when launched singly or in salvo, eventually proved devastatingly effective in combat. Ultimately, it played key roles in the defeat of the Nazis during the critical battles for Moscow, Leningrad, and other strategic Soviet positions.

The advent of World War II, which the Soviet Union came to call "The Great Patriotic War," proved the final death knell for the private Soviet rocket societies. It effectively cleared the way for a more formal, militarily-oriented approach to the study of astronautics and associated hardware design. During the course of the war, however, only minimal emphasis was placed on rocket-propelled weapons of any kind. As a result, the great intellectual and engineering talent that had been jelled by the societies and miscellaneous peripheral groups, virtually disappeared.

EXHIBIT ARTIFACT: Earth and Moon globes from the office of the great Soviet rocket engineer and rocket program director, Sergei Korolev.

SATELLITES AND PLANETARY PROBES

By 1954, the Soviet stage was set for one of the most significant technological feats in world history. The intercontinental ballistic missile concept had been accepted by the military establishment as viable; funding had been released for ICBM hardware development; propulsion system technology had moved ahead rapidly; early intermediate-range ballistic missile programs had proven the basic tenets of warhead delivery using inertially-guided rocket-propelled vehicles; control centers were under construction; and several launch facilities were either available or becoming available to accommodate the full-scale flight test programs then being scheduled.

While military interest in intercontinental range rockets generated the backing necessary to move designs from the drawing board to the launch pad, the more futuristic applications calling for these boosters to launch satellites and explore space were not forgotten. Proponents of such programs, including the influential Sergei Korolev and a broad band of academicians and various scientific academy members, eventually pushed the artificial satellite idea through the Kremlin and won approval for it during January of 1956.

Meetings to determine the physical characteristics and scientific objectives of the Soviet satellite, already referred to as *Sputnik* ("Traveler") now were held with considerable enthusiasm, and on September 11, 1956, it was announced by a Soviet contingent during an international conference in Barcelona, Spain, that the Soviet Union would launch an artificial satellite as part of the International Geophysical Year celebration.

On August 27, 1957, the famous Soviet newspaper *Pravda* ("Truth") reported successful testing of a 5,000 mile [8,047 km] range "intercontinental ballistic rocket" and successful detonations of "nuclear and thermonuclear weapons." These statements were virtually ignored in the West; their importance, however, was profound.

By mid-September, *Pravda* had run several additional stories alluding to forthcoming space-related events. One stated the Soviet Union was about to launch two Earth-circling satellites. This proclamation later was underscored by a follow-up story containing details for ham radio operators of the forthcoming satellite's radio-transmission frequencies. Oblivious to what was about to happen, Western observers again paid the claims only scant attention.

The dawning of October 4, 1957, changed this Western attitude... forever. Around the world, news announcements proclaimed that *Sputnik* 1, weighing 184.3 lb [83.6 kg], had been successfully placed in orbit by an R-7 launch vehicle; a manmade object had been rocketed into space and for the first time in history was circling the globe restrained only by the effects of gravity. Embarrassingly for the West, it made its presence known universally by broadcasting an electronic beep at 20.005 and 40.002 MHz that could be heard by anyone with the proper amateur radio equipment. Less than a month later, *Sputnik* 2 was rocketed aloft on November 3, 1957, weighing a stunning 1,121 lb [508.5 kg] and carrying a dog named *Laika* ("Husky"). *Laika* became the first living creature ever to orbit the Earth.

The Soviet accomplishments shattered the Western self-image of invulnerability and consequently demonstrated once and for all the Soviet resolve to become a leading world military and political power. Significantly, the accomplishment gave a stunning boost to Soviet self-esteem while rectifying, almost overnight, more than 25 years of frustration among Soviet space proponents.

With the death of Stalin in 1953 and the ascension of Nikita Khrushchev into power, the Soviet system changed dramatically and the country's position in world politics as an influential entity changed with it. Consequently, by the advent of *Sputnik*, the Soviet Union had created one of the world's most powerful military forces and had entered into service the *Shyster*, *Sandal*, and *Sapwood* missiles (these names are Western/NATO-assigned designators and are not utilized in the Soviet Union) — with the latter having intercontinental range and nuclear warhead capabilities.

The shock of *Sputnik* gave the U.S., and its military establishment in particular, much food for thought. Overnight, the reality and success of the Soviet rocket program became apparent not only to secretive intelligence communities, but to the public as well. The U.S. military's response was typically reactionary in nature, and by the end of the year, phrases such as "missile gap" and "space race" were heard across the country with considerable regularity.

EXHIBIT ARTIFACT: Sputnik 1, rocketed into orbit on October 4, 1957, was the world's first artificial satellite. Basically a metallic ball with two radio transmitters inside, it provided only minimal scientific data and had a relatively short service life. Regardless, as a propaganda tool it was one of the great coups of the 20th Century—and its effect on the world space community was everlasting.

Sputnik *1 satellite*
Full-scale replica
Launched: October 4, 1957
Launch vehicle: Sputnik
Weight: 184.3 lb [83.6 kg]

Two months after the stunning success of *Sputnik* 1, the first U.S. satellite, *Explorer* 1, weighing 31 lb [14 kg], was rocketed into orbit. Other U.S. satellites quickly followed, with a third *Sputnik* not attaining orbit until May 15, 1958. *Sputnik* 3 weighed an amazing 2,926 lb [1,327 kg] and was equipped with experiments for measuring radiation belts, corpuscular solar radiation, upper atmospheric density, ionosphere properties, magnetic fields, cosmic radiation, and micrometeorite impact effects.

Concurrent with the exploratory nature of the entire Soviet satellite program was a rapidly expanding interest in manned space flight. Accordingly, *Sputnik* satellites 4, 5, 6, 9, and 10 were launched during a ten-month period from May 15, 1960 to March 25, 1961, in order to test systems that potentially could be utilized to sustain a human being in space and effect a successful recovery. *Sputnik* 4 became the first of these research satellites, and with a weight of 9,990 lb [4,531 kg], was designed to explore, among other things, de-orbiting techniques. *Sputnik* 5, carrying the dogs *Belka* ("Squirrel") and *Strelka* ("Little Arrow"), followed and was successfully recovered, along with its four-legged explorers, some 24 hours after launch.

With the successes realized by most of the "manned" *Sputniks*, the Soviet space team, effectively under the skillful direction of the omnipresent Korolev, moved ahead with what was essentially a two-pronged space exploration program. One prong encompassed the development and launching of an extensive unmanned research and military satellite and space probe family; and the other encompassed the development and launching into orbit of the first manned space vehicles. The latter included a long-term plan to develop a permanent space station and eventually, to begin exploration of the cosmos.

From *Sputnik*, the unmanned satellite program expanded rapidly in the Soviet Union and today encompasses many hundreds of spacecraft. The following is a brief, selective overview of these vehicles and their mission objectives. ●

«Он был мал, этот самый первый искусственный спутник нашей старой планеты, но его звонкие позывные разнеслись по всем материкам и среди всех народов как воплощение дерзновенной мечты человечества.»

Сергей Королев

"It was small, that very first artificial satellite of our old planet, but its insistent signal resounded across the continents and among all their peoples like the realization of humanity's most daring dream."

Sergei Korolev

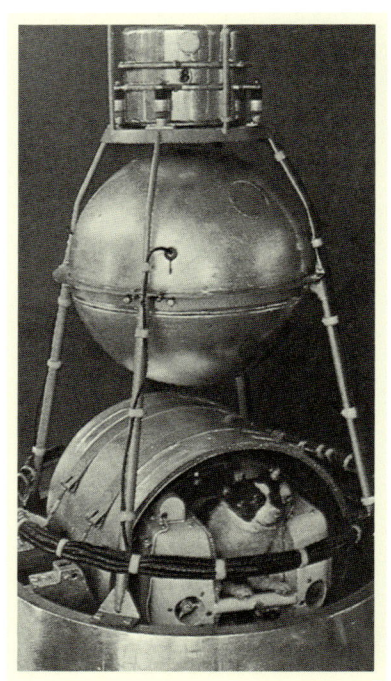

(right) Sputnik *2 carried a small dog,* Laika, *into space on what was to be the first Earth orbit ever by a living creature. Weighing over a half ton [508 kg],* Sputnik *2 provided Western intelligence bureaus with their first clear indication of Soviet booster lift capability. At the time, no U.S. booster could insert into orbit anything close to the weight of the first-generation Soviet satellite family.*

TERRESTRIAL SATELLITES

These represent the conventional family of Earth-circling satellites and are utilized primarily to accomplish research and/or military objectives.

Scientific:

Sputnik ("Traveler")—*Sputnik* 1 became the world's first artificial satellite when launched by the Soviet Union from Baikonur Cosmodrome into Earth orbit by a modified R-7 booster referred to as *Sputnik* (or SL-1 in the West) on October 4, 1957. With a diameter of 1 ft 10 in [.58 m], a weight of 184.3 lb [83.6 kg], and with only a modicum of onboard research equipment, it circled the globe until re-entering the atmosphere on January 4, 1958. Nine other *Sputniks* followed, these being designed to explore a variety of technical and scientific subjects. Highlights included *Sputnik* 5, which was the first spacecraft recovered from orbital flight and was one of several spacecraft in this series to carry live animals in conjunction with experiments to verify whether man could survive in space; and *Sputnik* 7 which, though unsuccessful, represented the first Soviet interplanetary probe to Venus. The *Sputnik* series resembled each other in name only and varied greatly in size, weight, and mission objectives. The last *Sputnik* launch took place on March 25, 1961.

Kosmos ("Cosmos")—A somewhat generic name assigned to a broad spectrum of satellites whose primary purpose is research in the physical, geological, biological, astronomical, meteorological and many other sciences. *Kosmos* experiments periodically involve not only those developed in the Soviet Union, but those developed in other countries, as well (the latter usually are referred to as the *Interkosmos* ["International Cosmos"] satellites). Many variations to the *Kosmos* satellite theme have been orbited and the designation continues in use as of this writing. Not all *Kosmos* satellites have been of a scientific nature, as the designation also is known to have been assigned to military reconnaissance and surveillance satellites. Well over two thousand *Kosmos* satellites and probes have been launched, with no apparent end in sight.

Polyot ("Flight")—The *Polyot* program is believed to have been an interim project designed to explore the feasibility of utilizing *Soyuz* ("Union") booster stages to propel a manned capsule to the Moon. Two *Polyot* test vehicles, equipped with large rocket engines to permit in-space maneuvering, were launched as satellites on November 1, 1963 and April 12, 1964. These satellites were declared by Soviet Premier Nikita Khrushchev to be the first maneuverable spacecraft ever to orbit the Earth.

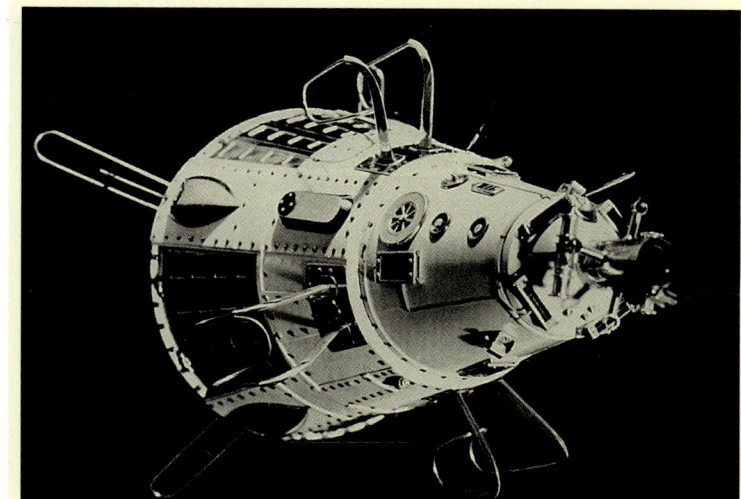

(above) *With a total weight of 2,926 lb [1,327 kg], Sputnik 3 stunned the American space community—whose heaviest satellite so far (Explorer 3) weighed less than 31 lb [14 kg].*

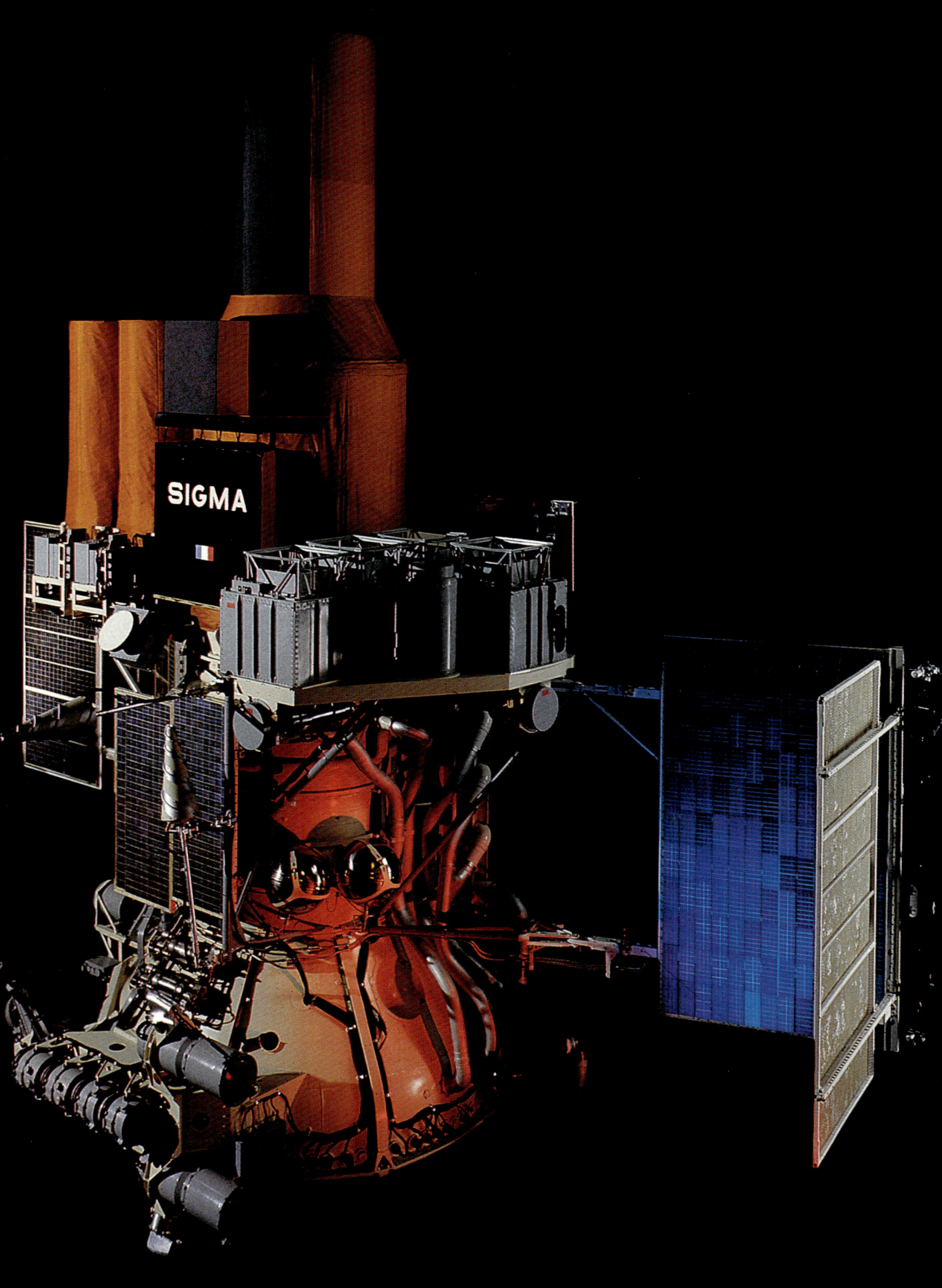

Automatic station Granat
Full-scale model
Launched: December 1, 1989
Launch vehicle: Proton
Weight: 9,700 lb [4,400 kg]

Elektron ("Electron")—Four *Elektron* satellites were launched in pairs into dissimilar orbits on January 30 and July 11, 1964, respectively, and were equipped to simultaneously study the inner and outer zones of the Earth's Van Allen radiation belts.

Proton ("Proton")—The four *Proton* satellites were designed to accommodate the study of high-energy cosmic and gamma rays and also the study of electrons of galactic origin. Weights varied from satellite to satellite, but the heaviest, *Proton* 4, grossed at approximately 37,485 lb [17,000 kg]. The first *Proton* was launched on July 16, 1965.

Prognoz ("Forecast")—A lengthy series of *Prognoz* satellites was launched in conjunction with the *Interball* program. With the exception of *Prognoz* 9 (which was optimized for cosmic radiation studies), these were designed to accommodate a wide variety of solar-related experiments including studies of the shock wave generated by the solar wind when it encounters the Earth's magnetosphere. The first *Prognoz* satellite was launched on April 14, 1972.

Astron/Granat ("Astronomer"/"Garnet")—Originally referred to as *Astron* 2, the 9,700 lb [4,400 kg] *Granat* satellite was equipped with an astronomical observatory optimized to study sources of X-ray and gamma ray emissions in space resulting from such phenomena as neutron stars, black holes, white dwarfs, and the remains of supernova eruptions. It carried, in addition to the telescope, seven gamma and X-ray detectors developed by Bulgarian, Dutch, French, and Soviet scientists. The satellite's primary observation tool, a Sigma X-ray telescope, was built by CNES—the French space agency. *Granat* was launched on December 1, 1989, and continues to generate data as of this writing.

Photon ("Photon")—The *Photon* materials science satellite is based on the *Vostok* design and weighs some 13,669 lb [6,200 kg]. At least one *Photon* launch per year is scheduled by the Soviets through the early 1990s—with the first having taken place during 1985. Each has a flight duration of up to 16 days. *Photon* 2 was the first spacecraft of Soviet origin to carry a foreign (French) commercial materials science package into space. *Photon* eventually is expected to be replaced as a research tool by the newer *Nika*-T spacecraft.

Oreol ("Aureole")—A series of three satellites designed to study the aurora borealis or "northern lights," and the upper latitudes of the Earth's atmosphere, containing the Van Allen belts.

(below) *The* Elektron *1 satellite was launched in conjunction with* Elektron *2 to simultaneously explore the inner and outer zones of the Earth's extensive Van Allen radiation belts. Both satellites were launched aboard the same booster, thus marking the first time a single vehicle was used by the Soviet Union to simultaneously insert two objects into orbit.*

EXHIBIT ARTIFACT: The Granat *orbital telescope is a dedicated gamma/X-ray observatory equipped with seven different detectors furnished by teams from Bulgaria, Holland, France, and the Soviet Union. The satellite's largest detector, a Sigma X-ray telescope, was built by the French space agency, CNES. Other* Granat *telescopes and detectors are sensitized to obtain low energy gamma and X-ray images from the depths of the universe.*

EXHIBIT ARTIFACT: The Bion *spacecraft is virtually identical to* Photon *and in turn, is derived from the original* Vostok *design which carried Yuri Gagarin into space in 1961. Several different* Bion *short-duration biological research missions have been launched since the first was sent into orbit during 1973. Always international in content,* Bion *satellite objectives tend to be broad with heavy emphasis on such items as motion sickness, reproduction, regeneration, immunology, and gravitational readaptation.*

Space apparatus Bion
Full-scale engineering model
First launched: October 31, 1973
Launch vehicle: Soyuz
Weight: 13,230 lb [6,000 kg]

Luna 3—Following launch on October 4, 1959, achieved a successful fly-by of the Moon on October 7, and in so doing, provided the first photo images ever of the Moon's far side—which is not visible from the Earth. Some 70 per cent of the Moon's far side was photographed, with images telemetered back to earth electronically. Weight was 613 lb [278 kg].

Zond 3 ("Probe")—Originally scheduled to be a Mars probe, this satellite, following launch on July 18, 1965, achieved a successful fly-by and photographed portions of the Moon's far side that had not been photographed by *Luna* 3. Weight was 2,095 lb [950 kg].

Soft-Landing Missions

Luna 4—Following launch on April 2, 1963, made an unsuccessful attempt to soft-land on the lunar surface. Missing the Moon resulted in a fly-by and eventual barycentric orbit around the Earth. Weight was 3,135 lb [1,422 kg].

Luna 5—Following launch on May 9, 1965, made an unsuccessful attempt to soft-land on the lunar surface. Failure resulted in a hard impact and the probe was destroyed. Weight was 3,255 lb [1,476 kg].

Luna 6—Following launch on June 8, 1965, made an unsuccessful attempt to soft-land on the lunar surface. Failure resulted in a fly-by and eventual solar orbit. Weight was 3,180 lb [1,442 kg].

Luna 7—Following launch on October 4, 1965, made an unsuccessful attempt to soft-land on the lunar surface. Failure resulted in a hard impact and the probe was destroyed. Weight was 3,321 lb [1,506 kg].

Luna 8—Following launch on December 3, 1965, made an unsuccessful attempt to soft-land on the lunar surface. Failure resulted in a hard impact and the probe was destroyed. Weight was 3,422 lb [1,552 kg].

Luna 9—Following launch on January 31, 1966, successfully soft-landed on the Moon and for three days returned panoramic television images of the lunar surface. This successful effort was a first in the history of lunar exploration and a major Cold War propaganda coup for the Soviet Union at the time. Weight was 3,490 lb [1,583 kg].

Luna 13—Following launch on December 21, 1966, successfully soft-landed on the Moon and returned panoramic television images and radiation and remote soil testing data. Weight was 3,572 lb [1,620 kg].

Orbital Missions

Luna 10—Following launch on March 31, 1966, successfully achieved lunar orbit as satellite for 56 days while telemetering research data back to Earth. Launch weight was 3,490 lb [1,583 kg].

Luna 11—Following launch on August 24, 1966, successfully achieved lunar orbit as satellite. It failed, however, to execute all of its mission objectives successfully and thus was referred to by the Soviets as a "spacecraft system and near-lunar space tester." Launch weight was 3,616 lb [1,640 kg].

Luna 12—Following launch on October 22, 1966, successfully achieved lunar orbit as satellite. Telemetered television images of lunar surface throughout orbital life. Launch weight was 3,585 lb [1,625 kg].

Luna 14—Following launch on April 7, 1968, successfully achieved lunar orbit as satellite. Like *Luna* 11, however, it failed to execute all of its mission objectives successfully. Weight was 3,561 lb [1,615 kg].

Luna 19—Following launch on September 28, 1971, successfully achieved lunar orbit as satellite and later was maneuvered into new orbit using onboard thruster unit. Weight was 12,300 lb [5,600 kg].

Luna 22—Following launch on May 29, 1974, successfully achieved lunar orbit as satellite. Was maneuvered extensively during its service life. Weight was 12,300 lb [5,600 kg].

Sample Return Missions

Luna 15—Following launch on July 13, 1969, failed to achieve soft lunar landing and was destroyed on impact. Weight was 12,570 lb [5,700 kg].

Luna 16—The success of the U.S. *Apollo* manned lunar exploration program during 1969 led the Soviet Union to continue its pursuit of lunar exploration using unmanned, remotely-controlled probes. In a successful attempt to acquire lunar surface core samples, *Luna* 16 was launched on September 12, 1970, and achieved a lunar soft-landing on September 17. The lander was equipped with an ascent/return-to-Earth stage which, following the gathering of several cores, rocketed from the Moon's surface, and on September 24, returned safely to Earth. This was the first ever successful remote collection of geological specimens from another celestial body. Weight was 12,624 lb [5,725 kg].

Luna 18—Following launch on September 2, 1971, failed to achieve soft lunar landing and was destroyed on impact. Weight was 12,570 lb [5,700 kg].

Luna 20—Following launch on February 14, 1972, successfully achieved lunar soft-landing. Echoing the success of *Luna* 16, lunar soil samples were collected and returned to Earth. Weight was 12,570 lb [5,702 kg].

Luna 23—Following launch on October 28, 1974, was damaged during lunar soft-landing attempt. Deep soil sample attempt failed and no sample was returned to Earth. Weight was 12,570 lb [5,700 kg].

Luna 24—Following launch on August 9, 1976, successfully achieved lunar soft-landing. Deep soil sample attempt and subsequent return of sample to Earth were both successful. This was the last *Luna* probe to go to the Moon. Weight was 12,570 lb [5,700 kg].

(below) Luna *24 Return Capsule, return to Earth sequence.*

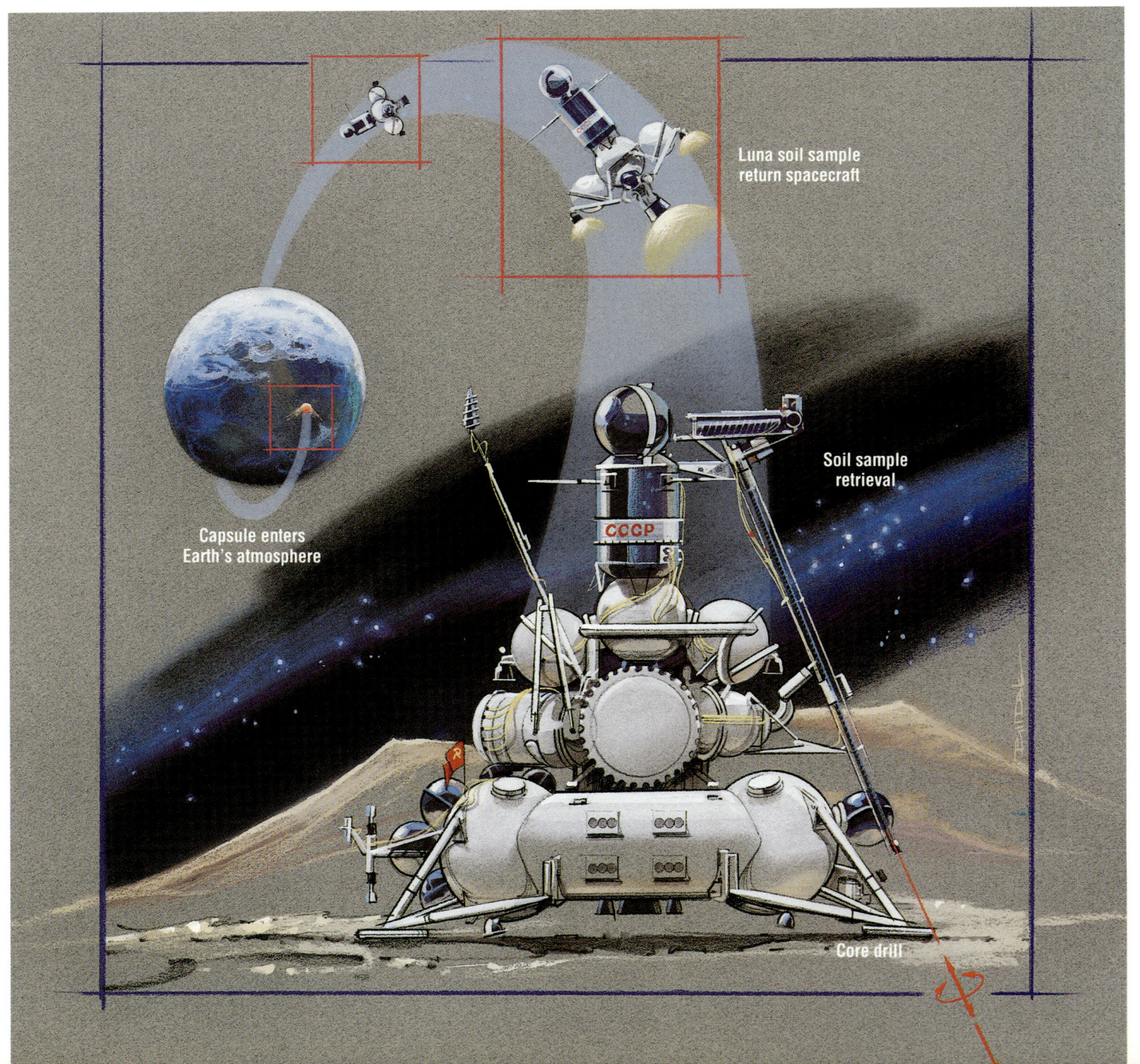

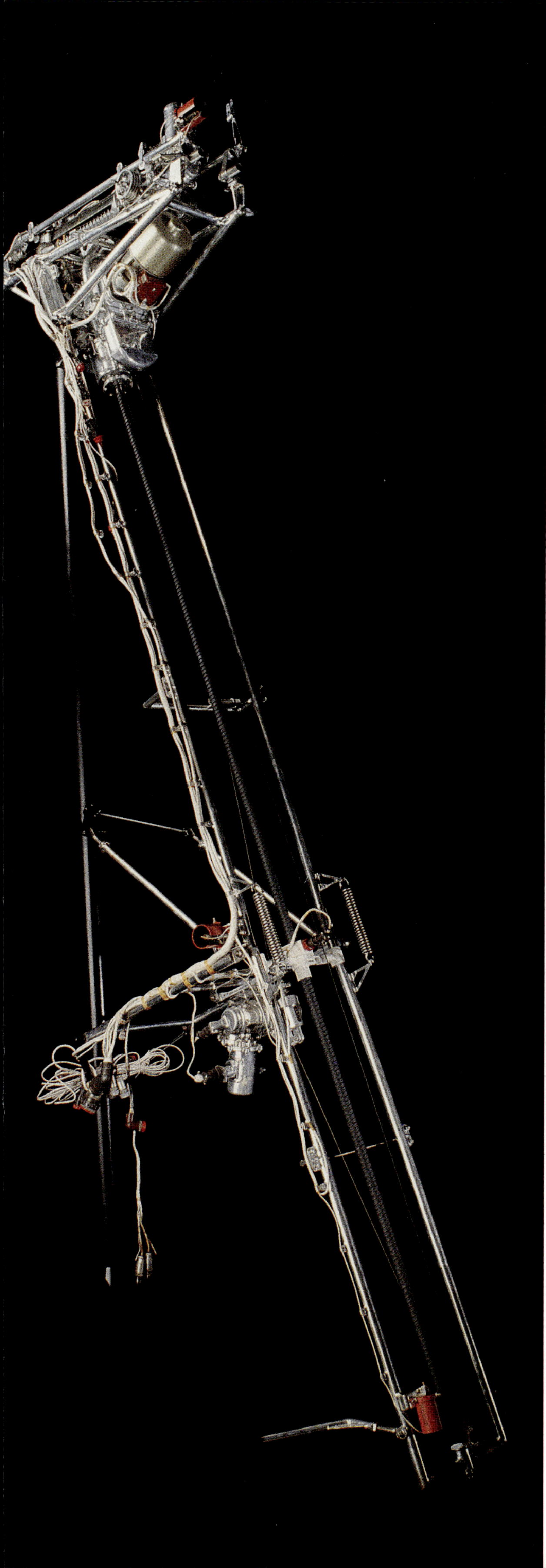

EXHIBIT ARTIFACTS: (left) *The* Luna *24 drill mechanism. This complex assembly was mounted on one side of the* Luna *probe and served to drill into the lunar surface and retrieve lengthy core samples. The latter, placed in a thin, tube-like plastic sleeve, was wrapped helically around a spool and inserted into the return capsule for transport back to Earth. This activity was remotely-controlled from Earth and was the last of its kind ever conducted.*

(below) *The* Luna *24 soil carrier tapes (top) and soil carrier and associated drum assembly. The latter served to accommodate the core samples following retrieval on the Moon's surface. The drum assembly, in turn, was inserted into the return capsule prior to being launched for the trip back to Earth.*

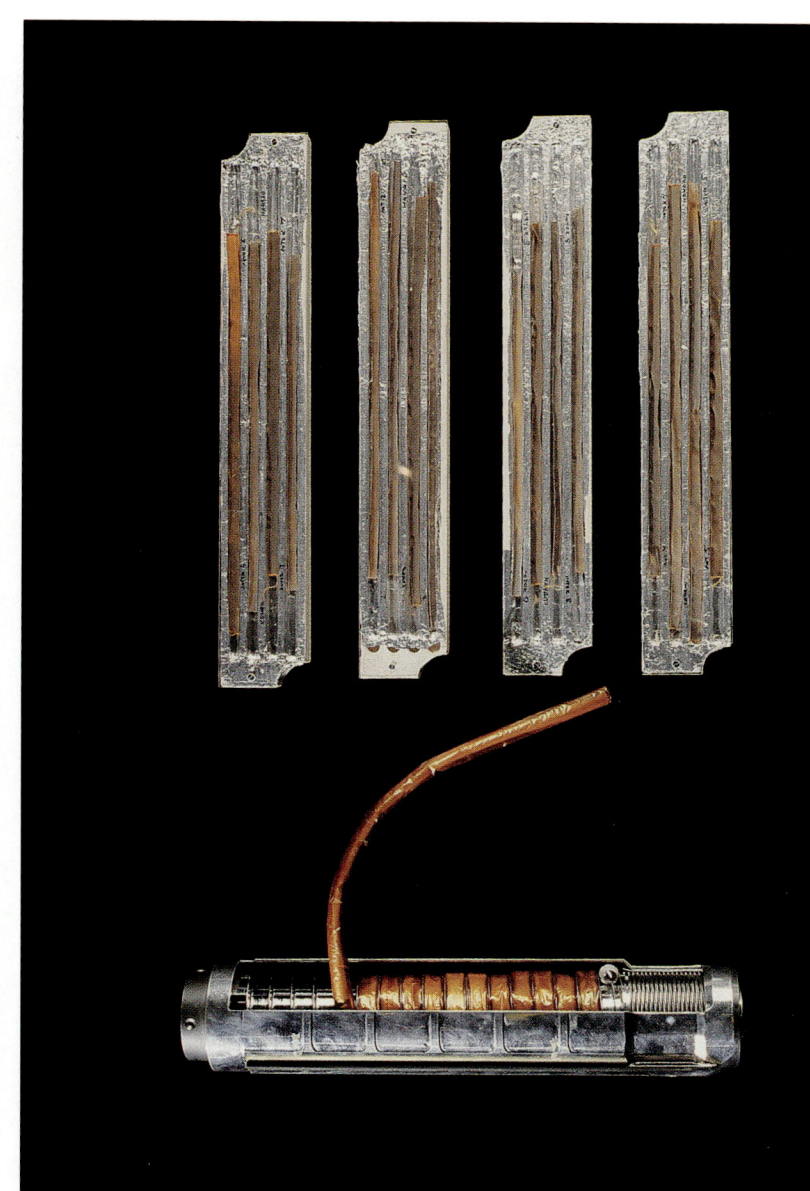

EXHIBIT ARTIFACT: A dynamic, full-scale working model of the Lunokhod 2 *remotely-controlled Moon rover which landed on the Moon's surface on January 16, 1973. This was the sixth successful Soviet lunar lander and the second surface rover. This vehicle represented an improved version of* Lunokhod 1. *It was faster as a result of more powerful electric motors, and it carried an extra vidicon television camera. Control of* Lunokhod 2, *following its arrival at the Moon's surface, was undertaken by a five-man crew at the Yevpatoria station in the Crimea.*

Lunokhod 2 *lunar rover*
Full-scale dynamic model
Launched: January 8, 1973
Landed: January 16, 1973
Launch vehicle: Proton
Weight: 1,852 lb [840 kg]

Unmanned Rover Missions

Luna 17/*Lunokhod* 1—Following launch on November 10, 1970, successfully achieved lunar soft-landing. First successful remotely-controlled lunar rover (*Lunokhod* 1) deployment following surface contact. *Lunokhod* 1 provided extensive television coverage during the course of its 6.5 mile [10.5 km]/ almost eleven month lunar surface tour.

Luna 21/*Lunokhod* 2—Following launch on January 8, 1973, successfully achieved lunar soft-landing. Second successful remotely-controlled lunar rover (*Lunokhod* 2) deployment following surface contact. Like its predecessor, *Lunokhod* 2 provided extensive television coverage during the course of its 23 mile [37 km]/four month lunar surface tour. ●

(right) Lunokhod *1 was the first of the Soviet lunar rovers to reach the Moon's surface when it made contact on November 17, 1970. It successfully gathered extensive data on lunar geology and sent back thousands of lunar images, such as this one, during the course of its several month sojourn on the lunar surface.*

(below) Lunokhod *sequence.*

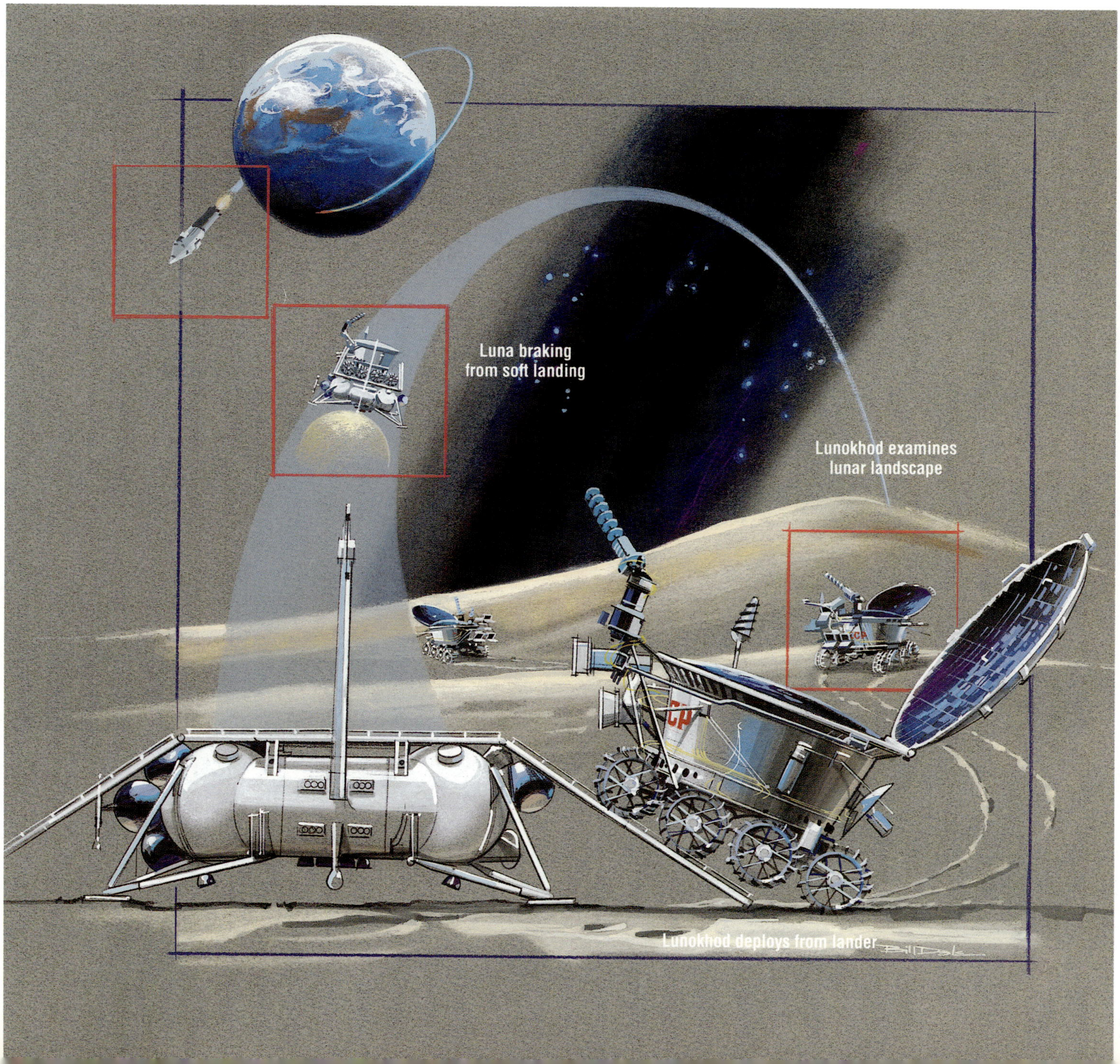

EXHIBIT ARTIFACT: The second-generation Venera *probe series, starting with Veneras 9 and 10, marked a new approach to the exploration of Venus. Independent orbiters and landers were launched together and upon arriving at the planet, separated in order to accomplish their respective mission objectives. This 1:10 scale model of* Venera *10 depicts both the orbiter and lander in their mated configuration.*

Automatic interplanetary station
Venera *10*
1:10 scale model
Launched: June 14, 1975
Launch vehicle: Proton
Landed: October 25, 1975
Weight: 11,098 lb [5,033 kg]

PLANETARY FLIGHTS

Following the early Soviet scientists' desires for interplanetary travel, the Soviet space program took an early interest in utilizing automatic or remotely-controlled probes to explore the known universe. The following describes the various unmanned probes that have been sent to orbit, impact, or land on the Earth's planetary neighbors for purposes of atmospheric and geologic research.

Venus

Venera ("Venus") 1 thru 16—The *Venera* space probe series was the Soviet Union's second attempt at planetary exploration. The first attempt to reach Venus began with the February 4, 1961 launch of the unsuccessful *Sputnik* 7, and was followed on February 12, 1961, by *Venera* 1. The latter was some 14,292,000 miles [23 million km] from Earth when contact was lost. Though a fly-by was accomplished successfully, there was no data transmission. *Venera* 2 followed on November 12, 1965, but also failed to maintain communication during the fly-by event. *Venera* 3, equipped with an atmospheric re-entry capsule, was launched on November 16, 1965, but like its predecessors, was unsuccessful except for the fact that it did impact the surface of Venus on March 1, 1966. Success with the *Venera* program finally was realized with *Venera* 4. Following launch on June 12, 1967, it soft-impacted the surface of Venus on October 18 and its research capsule successfully transmitted research data during the course of the descent through the Venusian atmosphere. *Veneras* 5, 6, 7, 8, 9, 10, 11, 12, 13, 14, 15, and 16 all were either partially or completely successful in terms of their entering the Venusian atmosphere or contacting the planet's surface; Venera 16, launched on June 7, 1983, was the last of the probe series. Two *Veneras*, 15 and 16, were equipped with radar systems permitting maps of the Venusian surface.

Zond ("Probe") 1—Optimized for a Venus fly-by, but failed when contact was lost approximately one month after launch on April 2, 1964.

EXHIBIT ARTIFACTS: (left) *A replica of the Venera 7 descent capsule. This was the first such device to reach the inhospitable Venusian surface and continue to transmit data to Earth. The capsule was lowered by parachute through the Venusian atmosphere. Data on surface temperatures and atmospheric pressure were among the several information bits returned to Earth bound receivers before the environment ended the probe's short life.*

(below) *A replica of the Venera 13 and 14 geological exploration drill. The Venera 13 lander arrived on the surface of Venus on March 1, 1982 and Venera 14 landed on March 5. Both landers included in their repertoire of experiments a soil probe that was capable of providing analytical data relating to soil chemical composition.*

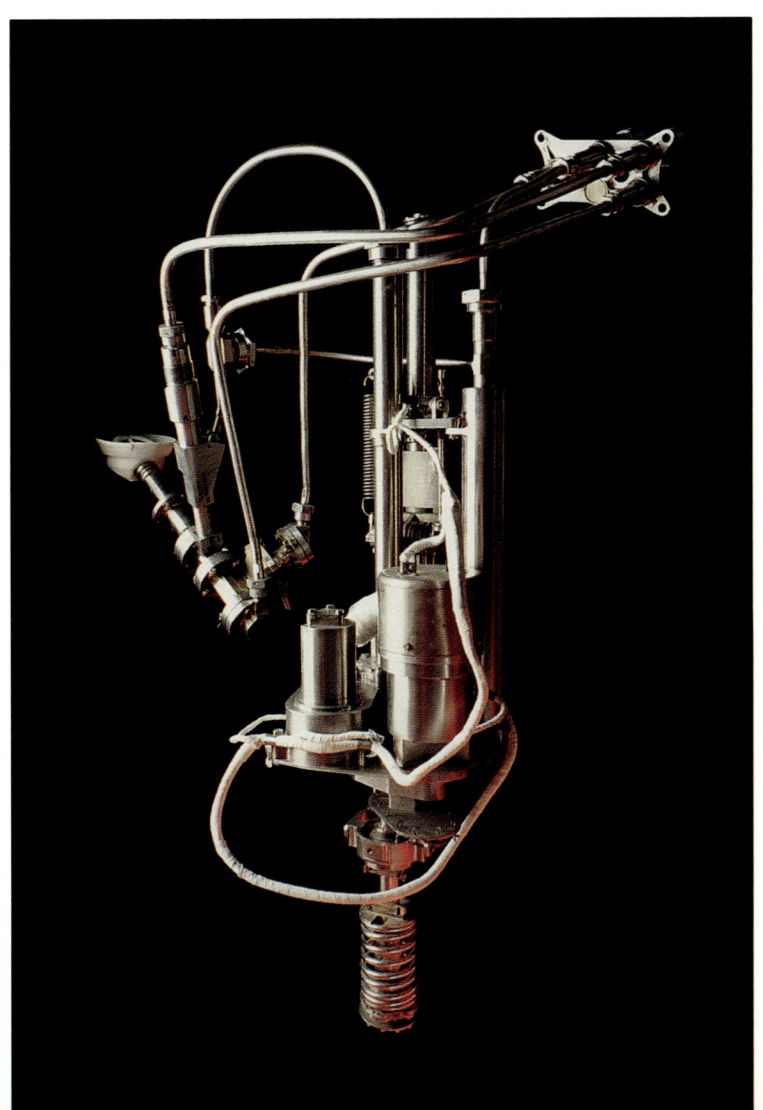

Automatic interplanetary station
Venera 7 lander
Full-scale engineering model
Launched: August 17, 1970
Launch vehicle: Molniya
Landed: December 15, 1970
Weight: 1,091 lb [495 kg]

Vegas 1 and 2—The realization that Halley's Comet would get close enough to fall within the launch envelopes of two *Veneras,* scheduled for late 1984 launch, led to a decision to move ahead with what became the *Vega* probes. These probes would be used to explore not only the original Venusian target, but Halley's Comet as well. The *Vega* probes were basically second-generation *Veneras* with increased strength and greater resistance to the extremely harsh Venusian atmosphere. The launch of *Vega* 1 took place on December 15, 1984, and on June 11, 1985, the probe's lander contacted the Venusian surface after deploying an instrumented balloon to drift in the atmosphere. In the interim, the probe itself continued on its way in pursuit of the comet. On March 6, 1986, *Vega* 1 came within 5,524 miles [8,890 km] of the nucleus of Halley's Comet, collecting data and transmitting digitized imagery of the noted space voyager back to Earth. *Vega* 2 was launched six days after *Vega* 1 on December 21, 1984, and was equally successful. It sent a probe to Venus on June 15, 1985, also deploying a balloon, and then came within 4,990 miles [8,030 km] of Halley's Comet's nucleus on March 9, 1986. Both *Vegas* transmitted digitized imagery and scientific data to Earth for scientific analysis.

(below) Vega *probe deployment sequence.*

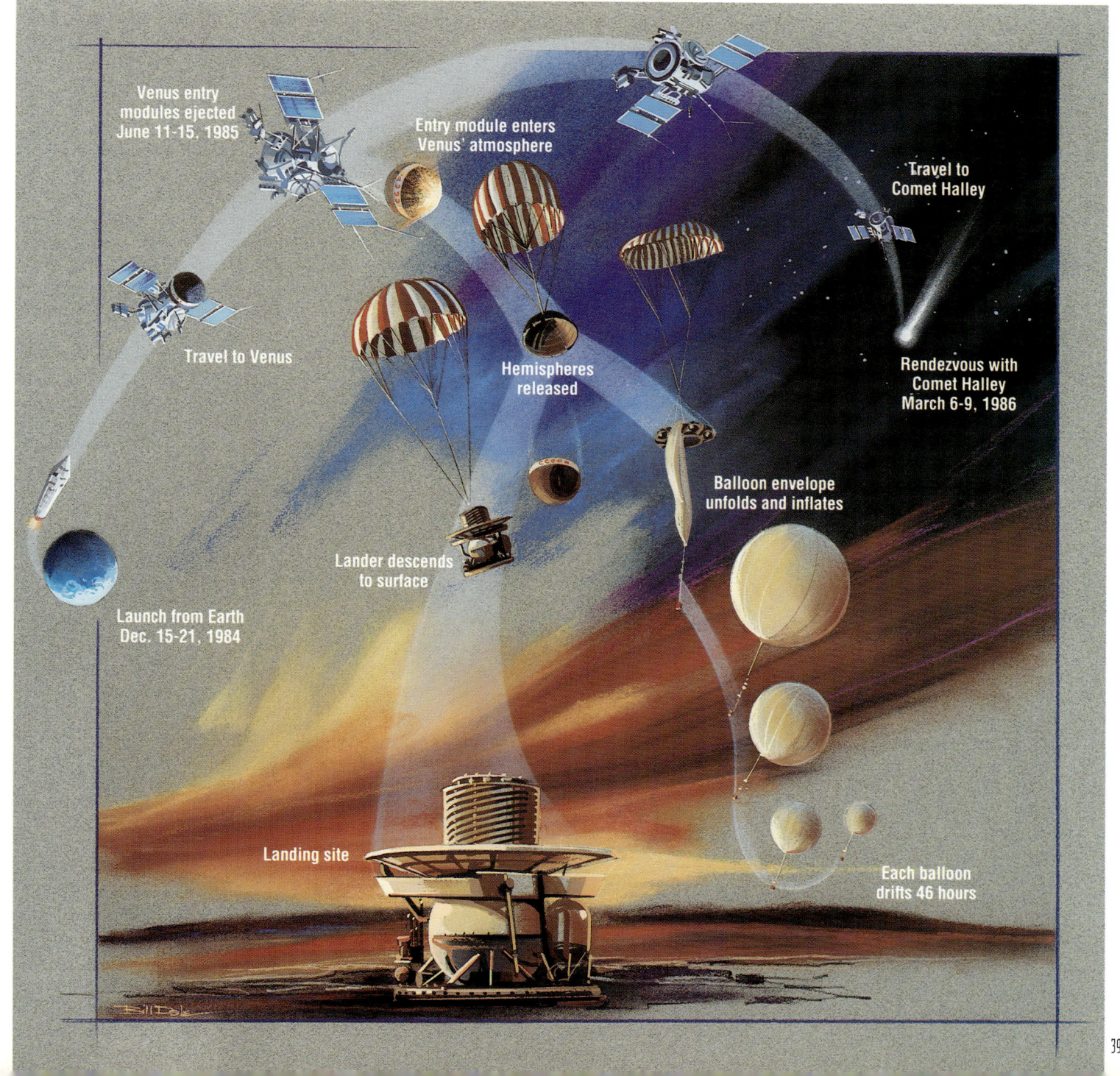

EXHIBIT ARTIFACT: Two Vega probes, modified from the second-generation Venera probe series, were launched on December 15 and December 21, 1984, respectively, with the intention not only of sending a research module to the surface of Venus, but also with the intent of sending a separate research module in pursuit of Halley's Comet. Both probes met with considerable success and consequently provided the world its first close-up images of Halley's Comet.

Automatic interplanetary station Vega
Full-scale engineering model
Launched: December 15 and 21, 1984
Launch vehicle: Proton

EXHIBIT ARTIFACT: Vega lander and descent sphere. Helium containers and other equipment can be seen mounted on top of the circular aerodisc. Below the latter is an aerodynamic stabilizing ring. Visible at the front left of the pressurized instrumentation sphere (inside) is a drill mechanism. The two chrome-like cylinders are part of the hygrometer and the large cylinder behind the instrumentation sphere is a gas chromatograph.

Vega *lander*
Landed: June 11 and 15, 1985

EXHIBIT ARTIFACT: Replica of the Vega balloon and instrument package that was injected into the Venusian atmosphere during the Vega 1 and Vega 2 missions launched on December 15 and December 21, 1984. Approaching the planet on June 9 and June 13, 1985, respectively, the balloons and their associated instrument packages were ejected to explore the Venusian atmosphere. The instrument package, referred to more accurately as an "aerostat," was suspended under the balloon and contained a number of gas, temperature, and related sensors.

Vega *aerostat*
Balloon diameter: 11 ft 6 in [3.5m]

Mars

Zond 2 ("Probe")—Though *Zond* 1, launched on April 2, 1964, was an unsuccessful *Venus* probe, *Zond* 2 became the earliest of the Soviet Mars probes. *Zond* 2 was launched on November 30, 1964, but also was lost when radio contact ended during April of 1965. Radar data indicated that *Zond* 2 got within 932 miles [1,500 km] of the Martian surface during its uncontrolled fly-by.

Mars 1 thru 7 ("Mars")—Following three launch and post-launch failures, *Mars* 1 was launched on November 1, 1962, on a trajectory to Mars until contact was lost five months later. The Soviets finally succeeded in getting a probe on its way to the red planet on May 19, 1971, when *Mars* 2 was safely injected into a viable Martian trajectory. On May 28, *Mars* 3 followed. Finally, on November 27, 1971, *Mars* 2 began its ascent into the Martian atmosphere. It was followed on December 2 by *Mars* 3. Unfortunately, at the time of both probes' arrivals, Mars was in the middle of the biggest dust storm ever recorded by Earth observers. *Mars* 2's surface probe eventually landed on Mars to become the first manmade object ever to reach the planet, but it failed to transmit any data. *Mars* 3 also landed, but transmitted data only for about 20 seconds before succumbing to system failure. Undaunted, the Soviets launched the *Mars* 4, 5, 6, and 7 probes, and eventually met with limited success with 5 and 6 following their respective arrivals on February 12 and March 12, 1974. *Mars* 5, like the unsuccessful *Mars* 4, was strictly an orbiter. *Mars* 6, (and the unsuccessful *Mars* 7), however, was a two-component configuration consisting of a descent capsule designed to carry experiments to the Martian surface, and a fly-by bus.

Phobos 1 thru 2 ("Phobos")—Two spacecraft, *Phobos* 1 and *Phobos* 2 were launched toward Mars during 1988, and came tantalizingly close to meeting their respective mission objectives—which were to investigate the larger of Mars' two small moons, Phobos. Unfortunately, *Phobos* 1 was lost en route as a result of an incorrect command sent from Earth. *Phobos* 2, however, after some six months in transit, eventually propelled itself into a Martian orbit. Following preliminary optical exploration of the planet, the probe was maneuvered into an orbit permitting it to access its primary objective—the moon Phobos. Unfortunately, just as the critical time came for the probe's closest Phobos pass, a control system computer malfunctioned and it became impossible to control. On April 14, 1989, it was declared lost by the Soviet Mars exploration team. ●

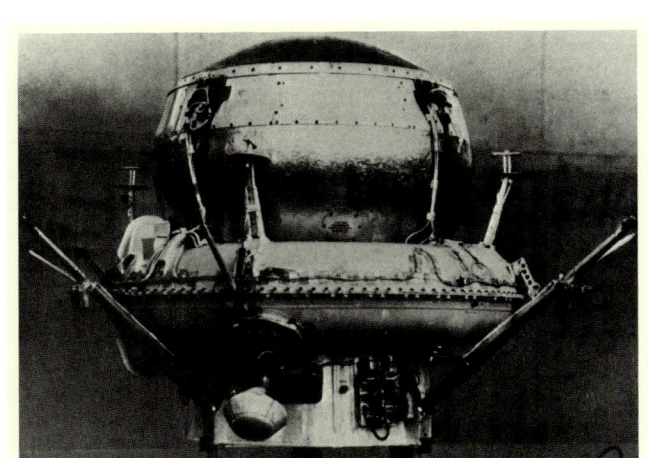

(left) Mars *2*/Mars *3 descent section without its associated heat shield. A capsule similar to this one became the first manmade object ever to physically contact the planet.*

EXHIBIT ARTIFACT: Mars 3 1:10 scale model. This spacecraft deposited a lander on the planet on December 2. Among its numerous experiments and sensors, it carried a French-manufactured Stereo 1 solar radio receiver. It was the second manmade object to land on Mars, being preceded by the Mars 2 probe which impacted the Martian surface on November 27.

Automatic interplanetary station
Mars 3
1:10 scale model
Launched: May 28, 1971
Launch vehicle: Proton
Weight: 10,253 lb [4,650 kg]

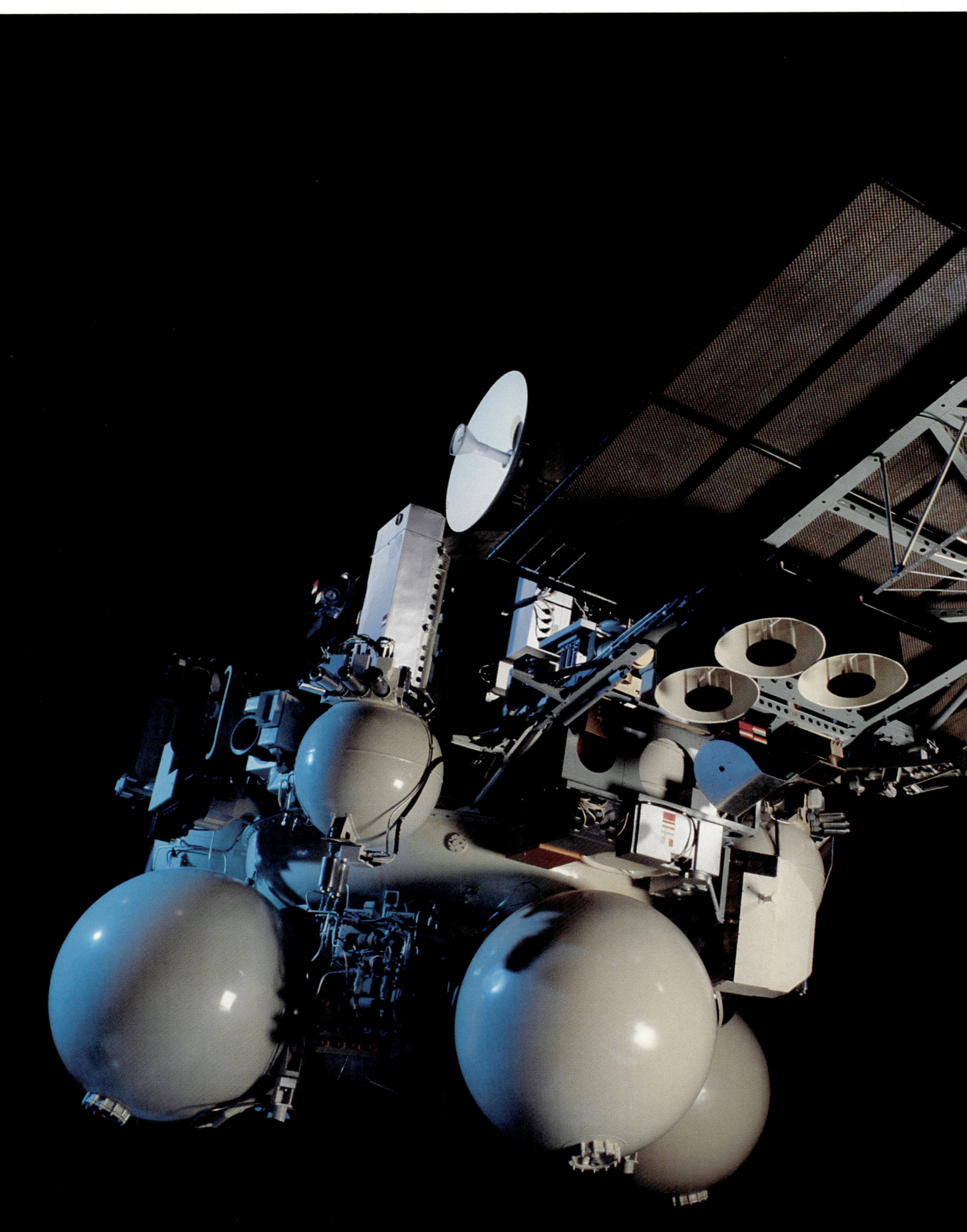

EXHIBIT ARTIFACT: Two Phobos *spacecraft were launched toward Mars during 1988 with the intent of renewing the Soviet exploration of that planet, that had effectively ended during 1974. Unfortunately, both* Phobos *probes failed on their way to Mars, including* Phobos *2 which was within days of a rendezvous with its namesake, the Martian moon, Phobos. Even though the missions were considered unsuccessful,* Phobos *2 had returned a number of high-quality, detailed images both of Mars and Phobos, thus providing scientists with significant insight into the red planet and its largest satellite.*

Automatic interplanetary station
Phobos
Full-scale engineering model
Launched: July 7 and 12, 1988
Launch vehicle: Proton
Weight: 13,715 lb [6,220 kg]

EXHIBIT ARTIFACT: *A full-scale model of the Oreol 1 ("Aureole") satellite. This was the first of three spacecraft launched as part of the Arcad (Arctic Auroral Density) project. A joint French and Soviet program, Arcad was optimized to study the aurora borealis or "northern lights" phenomenon and the upper atmosphere.*

Space apparatus
Oreol *1*
Full-scale model
Launched: December 27, 1971
Launch vehicle: Kosmos

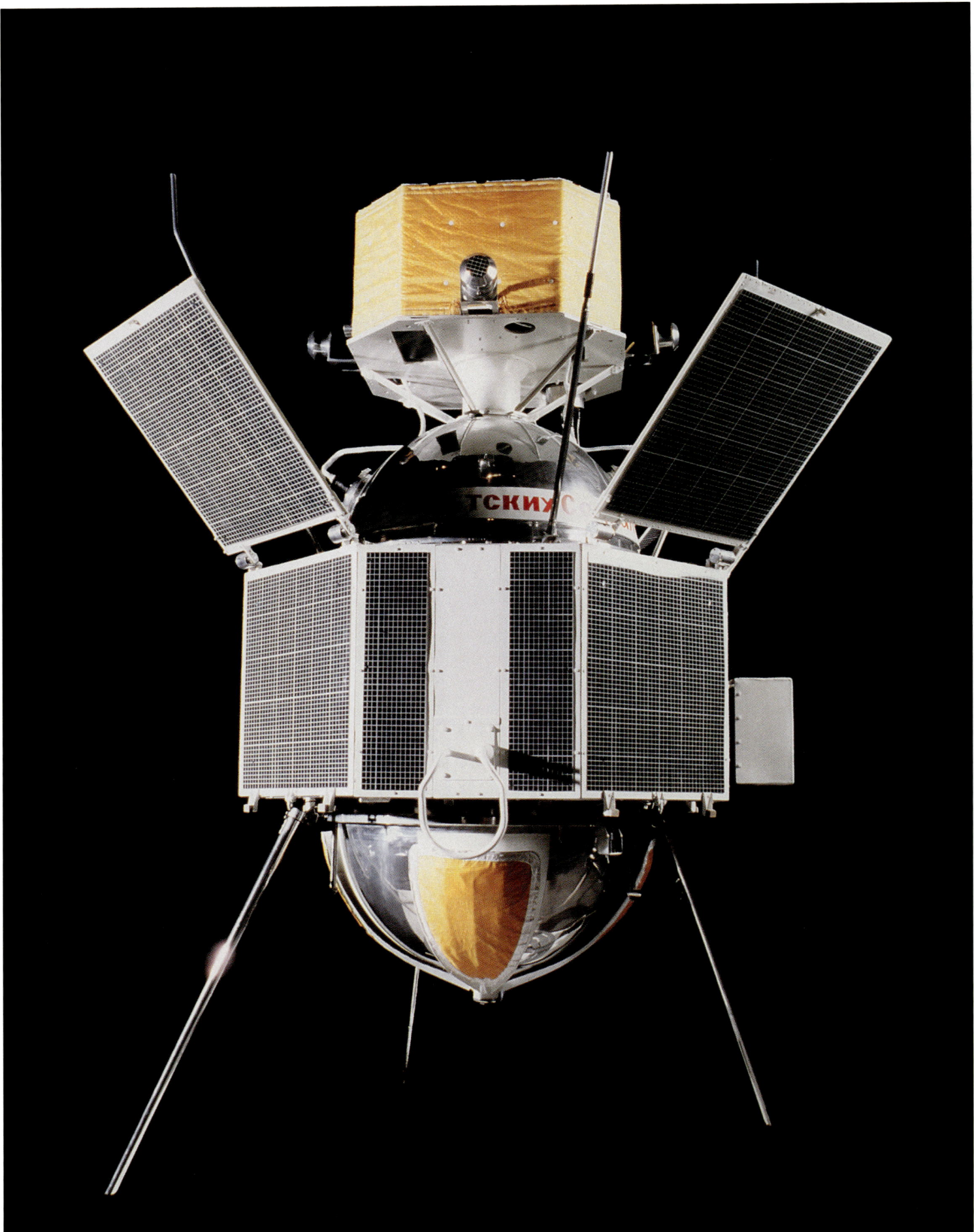

COSMONAUTS IN SPACE

Concurrent with the Soviet Union's 1954 decision to move ahead with an artificial satellite program, Sergei Korolev, by now the de facto leader of the entire Soviet rocketry and space program, began to look further into the future. Realizing that the payload capacity of the newly developed R-7 booster could, with modifications, be increased to accommodate the weight of a manned capsule and its associated life-support systems, he initiated a study that would follow the satellite with an even more spectacular achievement—the placing of a man into orbit. Tsiolkovsky's life-long dream of sending a man into space was finally about to be realized—nearly a quarter century after his death.

Korolev's commitment to manned space flight did not automatically assure its support in the Soviet space community. There were, in fact, opponents who argued strongly in favor of the more conservative unmanned satellite program. By 1958, however, a majority of the voices had been silenced, and by the end of November, a final decision had been made in favor of a manned spacecraft to complement the satellites. By the spring of 1959, design work and the construction of test hardware and a control center at Kaliningrad had moved ahead with great rapidity.

The new spacecraft, named *Vostok* ("East"), weighed a total of about 10,430 lb [4,730 kg] and consisted of two major components or modules. The manned component was a 7 ft 6 in [2.3 m] diameter metal sphere covered with heat-deflecting ablative material. Inside the sphere was a rocket-propelled ejection seat which exited through a circular hatch, a series of instrument panels and test packages, communications equipment, a television camera, and a viewing porthole equipped with a *Vzor* ("Visor") optical orientation device.

(above right) *The first cosmonauts were chosen during 1960. Members are shown in this team photograph which depicts in the front row (left to right), Pavel Popovich, Viktor Gorbatko, Yevgeny Khrunov, Yuri Gagarin, Sergei Korolev, Korolev's wife with Popovich's daughter, Karpov (the training director), Nikitin (the parachute trainer), and Dr. Fedorov (the team physician). The second row consists of (left to right) Alexei Leonov, Andrian Nikolayev, Mars Rafikov, Dmitri Zaikin, Boris Volynov, Gherman Titov, Grigori Nelyubov, Valery Bykovsky, and Georgi Shonin. The third row consists of (left to right) Valentin Filatyev, Ivan Anikeyev, and Pavel Belyayev.*

(middle) *The first man in space, Yuri Gagarin (left), is shown seated with the legendary Sergei Korolev, who had selected him for the historic mission.*

(right) *All Soviet manned space flights begin here, at the Baikonur Cosmodrome.*

EXHIBIT ARTIFACTS: (right) *A scale model of the* Vostok *booster. This booster was used for six manned and numerous unmanned space missions. Dependable, simple, and powerful, this basic booster configuration, derived from the R-7 ICBM, served as the foundation for the Soviet space program that exists today.*

(below) *A scale model of the* Vostok *manned spacecraft. The ball-shaped capsule contained the cosmonaut and his ejection seat, miscellaneous pieces of research equipment, observation ports, and atmospheric re-entry protection. The instrument module underneath served as the mounting bus for the life-support system oxygen and nitrogen bottles, miscellaneous instrumentation, and the retro-rocket system. Various communications antennas are visible at the top of the re-entry capsule and on the instrument module.*

(below right) *Used to ignite the engines of the* Soyuz *launch vehicles, this ignition key is inserted into the control console at the blockhouse, and turned around like a key in a car ignition.*

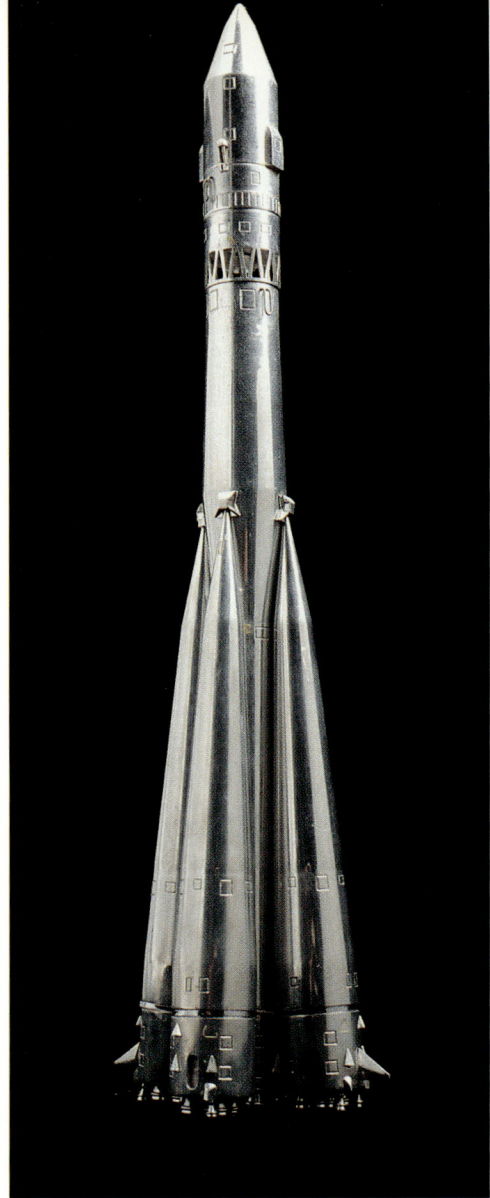

Vostok *rocket*
1:100 Scale model
Height: 126 ft [38.4 m]
Weight: 632,835 lb [287,000 kg]
First successful launch: January 2, 1959
(Luna *version*)

Vostok *spacecraft*
1:15 scale model
First manned launch: April 12, 1961
Launch vehicle: Vostok
Weight: 10,430 lb [4,730 kg]

The sphere was restrained by metal straps to an 8 ft 5 in [2.58 m] diameter ring permanently mounted atop a 7 ft 5 in [2.25 m] long equipment module. The latter served as the mounting device for the life-support system's high-pressure nitrogen and oxygen bottles (capable of providing a livable atmosphere for up to ten days in space); the electrical system's batteries; some test instrumentation; attitude thrusters for control in space; and perhaps most importantly, the retro-rocket that was required to slow the spacecraft for the de-orbiting procedure.

In space, the sphere and the equipment module were designed to separate from each other shortly after the retro-rocket had slowed the entire assembly for the re-entry process. The sphere, containing the cosmonaut, then would re-enter the atmosphere and descend to a predetermined altitude permitting the use of a descent parachute. Suspended from the latter, once the sphere had descended to an altitude of approximately 22,970 ft [7,000 m], the cosmonaut would eject and descend safely to the ground under his own parachute.

Five unmanned *Vostok* orbital test flights under the *Korabl-Sputnik* program were successfully completed between May 15, 1960 and March 25, 1961, before a final decision was made to rocket a man into space. During several of these pre-manned flights, dogs, anthropomorphic dummies, and a number of biological specimens were carried to see if space flight would cause any adverse physiological responses. Of the four *Korabl-Sputnik* flights equipped with a recovery capsule (mounted in the ejection seat), three were successfully retrieved for post-flight examination; no serious side effects on live specimens were noted.

While work on the first manned Soviet spacecraft proceeded at a rapid pace, the first team of Soviet spacemen, by now referred to as cosmonauts, also was being assembled. During early 1960, the first 20 cosmonaut-trainees were chosen, and these men were assembled shortly afterwards at the Khodynskoye Field facility for their initial exposure to the space program. By the beginning of 1961, several of the trainees either had been removed or had chosen to discontinue participation. Of the dozen remaining, two, Yuri A. Gagarin and Gherman S. Titov, had surfaced as the most likely candidates for the first manned *Vostok* mission.

(above) *The Soyuz booster was derived from the earlier* Vostok. *It differed primarily in the incorporation of a more powerful upper stage and the addition of a launch escape system to the payload shroud.*

(middle) *The business end of the immense* Vostok *booster. No less than 20 main exhaust nozzles are visible. The 12 smaller nozzles are verniers for steering and control. Total thrust at lift-off is over one million pounds [504,000 kg]. The four large booster units usually deplete their fuel loads approximately two minutes into flight and are ejected—with the central sustainer unit then becoming the main propulsive force. Variants with several upper stages have been developed and the booster remains in service in modified form.*

EXHIBIT ARTIFACTS: (below) Vostok *rockets are transported to their respective launch pads by railroad. Following arrival, large erector units mechanically maneuver the huge boosters to an upright position. They then are prepared for launch. Propellants are pumped into the launch vehicle's massive tanks following vertical placement on the launch pad.*

(right) *This simulation shows a* Vostok *booster at the moment of lift-off. All twenty main booster engines have ignited. Underneath the launch pad, the rocket exhaust is channeled away from the immediate area via an immense pit. At the same time, the large gantry assemblies have fallen away in order to give the rising booster and its cargo plenty of clearance.*

EXHIBIT ARTIFACTS: *Soviet newspapers and English-language Communist propaganda newspapers proclaim the early successes of the Soviet manned space flight program.*

(below) *The training suit of Vladimir Dzhanibekov from the Soyuz 27-Salyut 6 mission of 1978.*

Khodynskoye Field eventually proved a temporary cosmonaut training center. Within a short while after the initial twenty cosmonauts were chosen, a more sophisticated and considerably better equipped facility was under construction near Moscow. This eventually became more formally referred to as *Zvezdny Gorodok* or "Star City," and was to become the site of the cosmonaut-dedicated Gagarin Training Center. In Star City, all cosmonaut training, both physical and intellectual, took place, and appropriate spacecraft mock-ups and training tools were provided. Additionally, an autonomous community that included housing for the cosmonauts' families, was created.

On April 10, 1961, Gagarin and Titov were informed that Gagarin would become the first Soviet cosmonaut. Titov would serve as his back-up. *Vostok* 1, by then, was in final preparation for launching at the Baikonur Cosmodrome.

On April 12, 1961, at 9:07 a.m. Moscow time, Yuri Gagarin and *Vostok* 1 lifted-off from Baikonur Cosmodrome. One hour and 48 minutes later, Gagarin was back on the ground, having become the first human in history to orbit the Earth. The mission was a complete success.

Vostok 2, with cosmonaut Gherman Titov onboard, served as yet another bold step forward. Launched on August 6, 1961, it remained in space for no less than 25 hours. Titov returned safely to Earth, but as a result of his experiences with space sickness during the course of the mission, he was temporarily grounded.

Andrian G. Nikolayev became the third Soviet cosmonaut when he was launched aboard *Vostok* 3 on August 11, 1962. The following day, Pavel R. Popovich became number four when he was launched aboard *Vostok* 4. Though the two spacemen came within 4 miles [6.5 km] of each other, their *Vostoks'* limited maneuverability prevented the consummation of any plans calling for an actual rendezvous. Regardless, the fact that the Soviet Union had orbited two cosmonauts simultaneously was an act of impressive dimensions. It underscored the country's confidence in its launch vehicles and technology and perhaps most importantly, it alluded to a significant long-term capability that eventually would lead to space stations and long-duration space missions.

«Юрий Гагарин поднимается к лифту стартовых ферм, и в этот миг обжигает меня очень простая, но новая, ранее неведомая мысль: «Да ведь это уже не тренировка! Ведь это — реальный полет!...»

А было это 25 лет назад и незаметно превратилось в историю.»

Герман Титов

"Yuri Gagarin walks up to the elevator at the launching pads, and at that very moment I am stunned by a very simple thought, which had never before entered my mind: Hey, this isn't training! This is the real thing!.. And this happened 25 years ago and is now just history."

Gherman Titov

(above) *Soyuz booster launches are very dramatic. All manned missions have taken place from the Baikonur Cosmodrome some 1,500 miles [2,414 km] southeast of Moscow.*

(right) *All Vostok manned space capsules were equipped with an ejection seat to permit cosmonaut egress prior to capsule ground contact. Following a rocket-powered ejection from the capsule, the cosmonaut resorted to a conventional parachute for the final segment of his return trip to Earth.*

On June 14, 1963, *Vostok* 5 was launched with cosmonaut Valery F. Bykovsky onboard. Remaining in space for some five days, this became the longest one-man space flight ever.

To this point, only men had been rocketed into space. In an effort to correct this inequity, and at the same time to create a major public relations coup, five women were recruited to join the cosmonaut program. Eventually, Valentina Tereshkova (rumored to have been picked at the last moment over the prime woman cosmonaut, Irina Soloveva) was assigned as the first woman space voyager, and on June 16, 1963, was rocketed into space aboard *Vostok* 6. She remained in space for nearly three days before safely returning.

By mid-1963, the original *Vostok* spacecraft had outlived its usefulness. A larger and more capable system now was required, and conveniently, the advent of a developed and more powerful version of the original R-7 gave the Korolev team a considerable number of options.

Though a number of studies for a *Vostok* replacement had been conducted, the two most important, *Vostok* Zh and what had become known as the *Soyuz* complex spacecraft, remained paper exercises. The incentive to expedite their completion appeared for the time at least, lacking. Complicating this was a decision to redirect much of the original *Soyuz* program and consequently to redesign a significant portion of the hardware. The latter were end products of the on-going manned Moon landing program and the Soviet space bureaucracy's inability to focus on a definite approach to the Moon landing problem.

In the interim, in order for the Soviets to maintain a presence in space, a decision was made to forge ahead with what became known as the *Voskhod* ("Sunrise") spacecraft. This *Vostok* successor externally differed little from its predecessor but was capable of accommodating up to three cosmonauts. Other noteworthy changes included in *Voskhod* 1, a back-up retro-rocket; and in *Voskhod* 2, an extendible airlock. The latter permitted a cosmonaut to undertake space walks (extra-vehicular activity; EVA) and conduct experiments outside the capsule while it was in orbit.

Though the changes appeared few, the added systems and refinements had increased the weight of the *Voskhod* capsule by nearly 2,000 lb [907 kg] over that of *Vostok*. Accordingly, the new booster, with its greatly increased payload capability, was required to get it into space. Additionally and importantly,

(above left) *Valentina Tereshkova became the first woman to orbit the Earth when launched aboard* Vostok *6 on June 16, 1963.*

(left) *A full-scale* Vostok *spacecraft attached to its orbital stage. The ball-shaped re-entry capsule was separated from the instrument module prior to re-entry and after the release of the large attachment straps.*

EXHIBIT ARTIFACT: *Yuri Gagarin's sunglasses and miscellaneous cosmonaut memorabilia.*

The three major Soviet cosmodromes from which satellites, probes, and manned spacecraft are launched include Plesetsk, Kapustin Yar, and Baikonur. Though Baikonur is the oldest and most famous, Plesetsk is the busiest. More launches have occurred at Plesetsk than at any other launch center in the world. Star City, near Moscow, is the location of the Gagarin Cosmonaut Training Center.

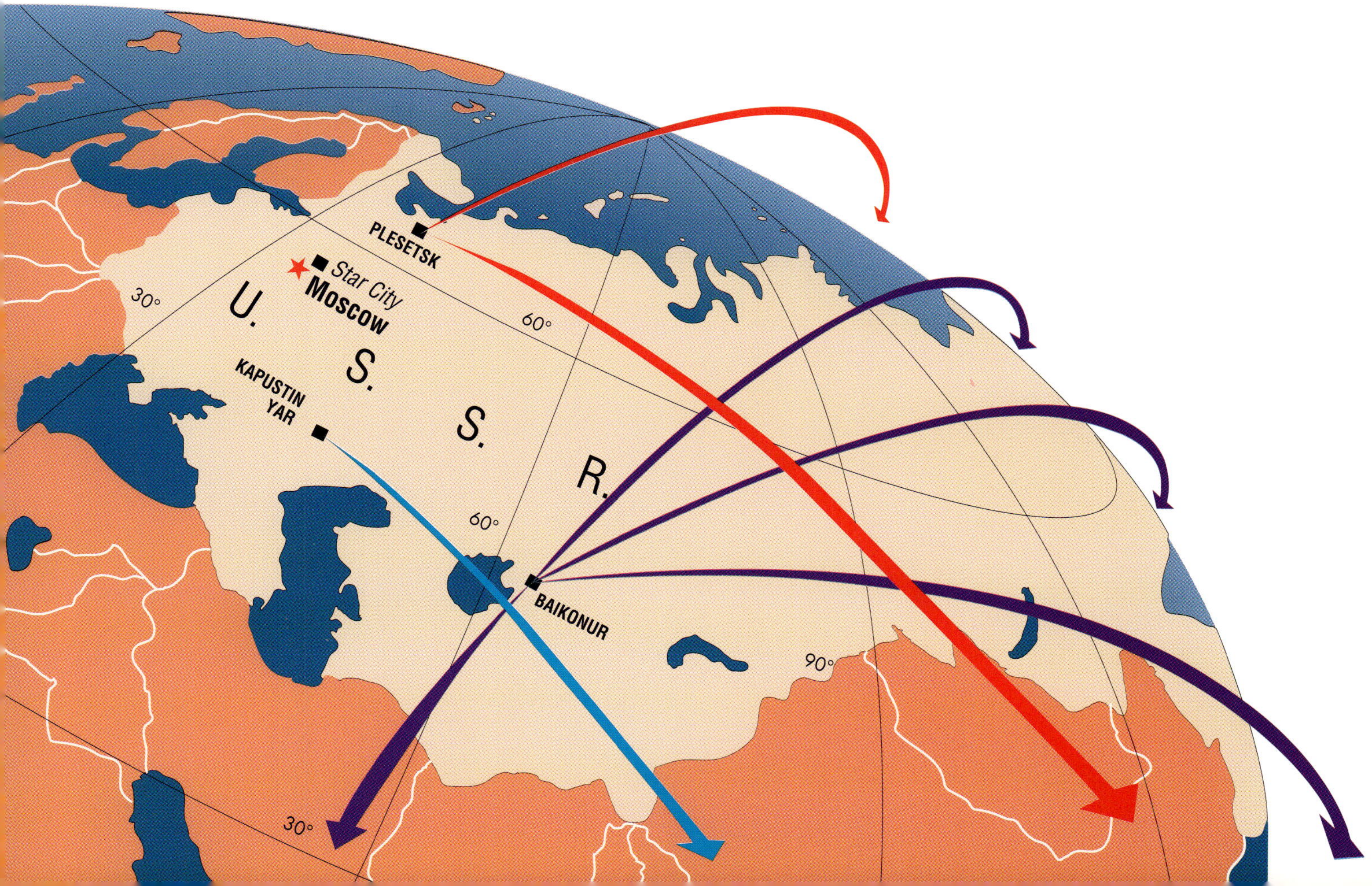

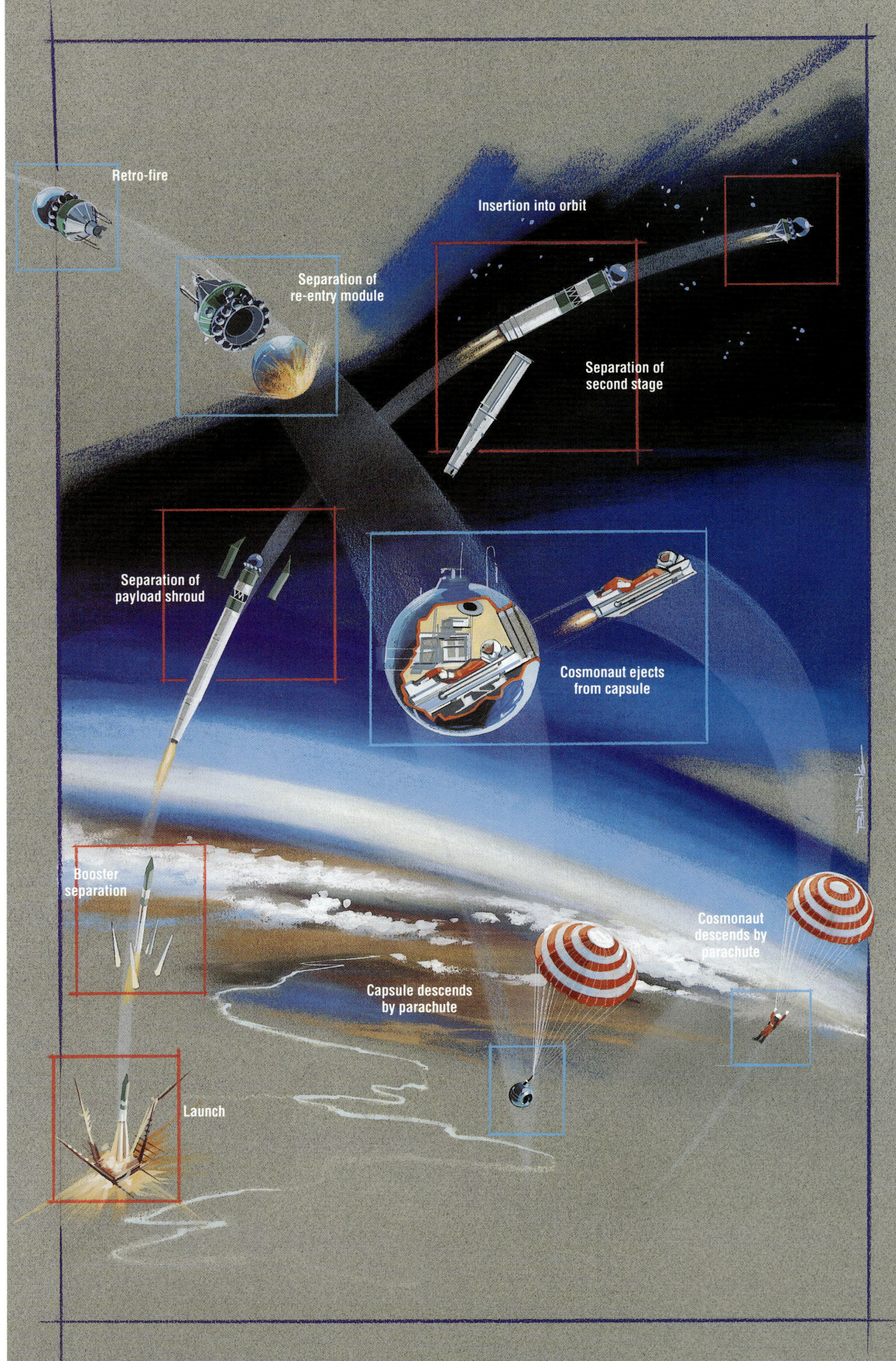

EXHIBIT ARTIFACT: A Vostok-type return capsule and its deployed, post-re-entry parachute and associated harness assembly. This particular capsule carried a payload of live animals into space during a 1983 mission—some 22 years after Yuri Gagarin's historic space flight in a similar capsule. Derivatives of the Vostok capsule remain in use as materials processing laboratory, biological laboratory, photo reconnaissance, and miscellaneous military payload transports.

Vostok-derivative return capsule with parachute
First launched: May 15, 1960
Launch vehicle: Vostok
Weight: 5,424 lb [2,460 kg]

(left) *Vostok launch and re-entry sequence. Particularly noteworthy is the cosmonaut ejecting from the ball-shaped capsule, and landing with a separate parachute.*

Voskhod represented the first Soviet manned spacecraft that permitted the cosmonauts to return to Earth without having to use an ejection seat first.

The first *Voskhod* launch (and the first launch of a spacecraft with a three-member crew), with cosmonauts Vladimir Komarov, Boris Yegorov, and Konstantin Feoktistov aboard, took place at Baikonur Cosmodrome without incident on October 12, 1964. It remained in orbit for just over 24 hours and returned to Earth for a normal recovery. It was followed, on March 18, 1965, by *Voskhod* 2, manned by Pavel Belyayev and Alexei Leonov. This flight proved decidedly more spectacular as it set the stage for what became the world's first space walk. Approximately at the end of the first orbit, cosmonaut Leonov opened the outer hatch of the special airlock and stepped into space — and history.

He would later recall, "I climbed out of the hatch, pushed myself away from it gently, moving farther and farther away from the ship. The lifeline that connected me with the ship stretched to its full length and then my movement away from the ship ceased. The slight effort I had made in detaching myself from the ship had caused it to move slightly and I saw our wonderful spaceship turning slowly before my eyes. I expected to see sharp contrasts of light and shadow, but there was nothing of this kind. The parts of the ship in the shadow were illuminated well enough by the sun's rays reflected from the Earth. I pulled the lifeline slightly and started slowly moving away again and began to move gradually away from the spaceship, turning about my transverse axis. I saw the universe in all its grandeur. The view of untwinkling stars on a velvet black background . . . was followed by views of the Earth . . . I recognized the Volga, the mountain range of the Urals . . . as though I were swimming over a vast colorful map . . . I knew that it was impossible to stop my rotation by any movements . . . so I merely waited for the rotation to slow down when the lifeline became taut by twisting. And soon the speed of my movement gradually decreased . . .

"Some time later I made a pretty strong pull at the lifeline and was forced to protect myself from the spaceship which started moving swiftly toward me. My first thought was not to strike the spaceship with the visor of my helmet. So, as it flew towards me, I softened the blow with my hands. This proved very easy to do . . .

EXHIBIT ARTIFACT: (right) *Six of these RD-253 rocket engines are employed in the first stage of the* Proton *launch vehicle, producing a total of 2,024,190 lb [918,000 kg] of thrust. This updated version of the original design (which was first flown in 1965) features a side-mounted turbo-pump assembly.*

(above right) *As in the West, Soviet cosmonauts train for space missions by working on full-scale mock-ups suspended underwater in immense tanks. The buoyancy of water suspension is quite similar to that of space and permits realistic training to be undertaken.*

(right) *A* Voskhod *capsule being prepared for flight. Basically a modified* Vostok, *it was equipped to carry up to three cosmonauts. It differed in having a retro-rocket system added to the parachute pack for landing; not requiring the cosmonauts to egress by ejection seat prior to ground contact; and in the case of* Vostok *2, having an airlock assembly to permit space walks.*

Rocket engine
RD-253
Oxidizer: Nitrogen tetroxide
Fuel: UDMH
Thrust: 337,365 lb [153,000 kg] - sea level

EXHIBIT ARTIFACT: Full-scale replica of a Kosmos 110 payload container for the dogs Veterok *("Breeze") and* Ugolek *("Small Piece of Coal"). The two dogs were in side-by-side containers mounted in the spacecraft and were orbited by a Soyuz booster. The dogs remained in orbit for 22 days and upon returning safely to Earth, held the record for the longest duration and highest orbit of any living creatures at the time. Experiments related to radiation exposure and weightlessness were conducted. A television camera was provided to permit scientific observation of the two animals during the course of the mission.*

Though the success of *Apollo* effectively ended the original *Soyuz* Moon mission, it did not end *Soyuz*. *Soyuz*, in fact, was simply reoriented and found to be ideal for use as a ferry between the Earth and what now was becoming a strong Soviet thrust in the direction of space station development.

At the time of the first *Soyuz* launch on April 23, 1967, with veteran cosmonaut Vladimir Komarov onboard, the program remained in a state of indecision because of the difficulties then confronting the various elements of the Moon exploration team. Regardless, *Soyuz* 1 was optimized as a testbed for the lunar-configured spacecraft and its basic mission objectives were to determine the viability of the design for lengthy stays in space.

Though the launch of *Soyuz* 1 (utilizing the new *Soyuz* booster) went smoothly, shortly after the vehicle entered orbit, difficulties apparently arose. Available information suggests that wings containing the solar cells which were to supply electrical power for onboard systems did not deploy properly, and that within hours of entering Earth orbit, Komarov was in trouble. Several re-entry attempts failed, and apparently on the third, matters went from bad to worse as the ball-shaped capsule began its fiery trip back to Earth. Probably due to an unstable re-entry, the spacecraft's parachute lines fouled during the descent. By the time rescuers reached Komarov in the charred capsule, he was dead—the victim of extremely high impact forces.

The failure of this first *Soyuz* mission led to several interim, unmanned tests of improved *Soyuz* configurations under the *Kosmos* satellite banner. It was not until October 25, 1968, that another *Soyuz* flight was attempted. This spacecraft was unmanned, but it was followed into space on October 26 by *Soyuz* 3 carrying cosmonaut Georgi Beregovoi. Beregovoi later maneuvered his spacecraft to within 656 ft [200 m] of *Soyuz* 2, thus setting the stage for the later spacecraft matings that would prove so critical to future space station activities.

On January 14 and January 15, 1969, *Soyuz* 4 and *Soyuz* 5 respectively, were rocketed into Earth orbit aboard their *Soyuz* boosters. *Soyuz* 4 was manned by Vladimir Shatalov and *Soyuz* 5 was manned by Boris Volynov, Alexei Yeliseyev, and Yevgeny Khrunov. This proved to be a memorable mission, for on January 16, the two spacecraft and their four crewmen completed the first-ever successful docking of two manned vehicles in space.

The docking was followed by an EVA-type (space walk) crew transfer wherein Khrunov and Yeliseyev, after donning spacesuits and crawling out of a special hatch into space, moved from *Soyuz* 5 to *Soyuz* 4 and returned to Earth in the latter. By hand-walking between the two *Soyuz* spacecraft, it became possible for the cosmonauts to change spacecrafts. Both spacecraft returned to Earth safely, with *Soyuz* 4 landing on January 17 and *Soyuz* 5 landing on January 18.

The next Soviet space spectacular began on October 11, 1969, when *Soyuz* 6 was rocketed into orbit carrying Georgi Shonin and Valery Kubasov. In turn, on October 12, *Soyuz* 7 was rocketed into orbit carrying Anatoly Filipchenko, Vladislav Volkov, and Viktor Gorbatko. On October 13, a third *Soyuz* was propelled into space, this one carrying Vladimir Shatalov and Alexei Yeliseyev—who had both been on the two previous *Soyuz* missions.

The objectives in having three spacecraft in flight at the same time were manifold, but perhaps most importantly, the Soviet space team wanted to explore the difficulties entailed in controlling and maneuvering simultaneously three bodies under the strong influence of orbital dynamics. All three spacecraft were maneuvered at the same time and two, *Soyuz* 7 and *Soyuz* 8, came close enough to permit reciprocal photography.

On October 16, 1969, *Soyuz* 6 returned to Earth without difficulty, and on October 17, *Soyuz* 7 followed. *Soyuz* 8 landed safely the next day.

The final flight of the first major phase of the *Soyuz* program took place with the launch of *Soyuz* 9 on June 1, 1970. The crew consisted of Andrian Nikolayev and Vitaly Sevastyanov. The mission's singular biological objective, other than a series of miscellaneous in-flight experiments and photography from space, was to study the effects of long-term weightlessness on the human body. The resulting mission, which lasted just over 17 days and ended without incident, proved fruitful, but a self-induced spinning of the spacecraft for stabilization purposes generated mixed reviews of the resulting biological effects. ●

EXHIBIT ARTIFACT: An RD-301 rocket engine. This is an experimental propulsion unit developed during the 1970s to test the feasibility of using so-called "exotic" propellants. No missions actually were flown utilizing this engine, but valuable experience was gained.

SPACE STATIONS

With the end of the *Soyuz* 9 mission and the earlier decision to terminate all projects related to transporting a cosmonaut to the Moon's surface, the Soviet Union began a major reassessment of its manned space program and all the associated philosophical, technical, military, biological, and political ramifications it entailed. A subtle, but strong hint of realignment had begun to take shape as early as the end of the *Vostok* program during 1963 when it became evident that the Soviet space bureaucracy lacked a strong and unequivocating sense of direction. Though Korolev felt confident that the next major step should be manned exploration of the lunar surface, a significant number of less influential proponents disagreed and quietly pursued goals of a less lofty nature. Though these would have little short-term effect on extant Soviet space policy, their long-term effect would eventually generate a major shift in space exploration philosophy.

Foremost among the non-lunar exploration options was the development of a true space station and the eventual insertion into Earth orbit of a facility that could accommodate human occupation for long periods of time. Shortly after the triumph of *Apollo* 11 wherein Neil Armstrong became the first man to walk on the Moon, a massive re-analysis of Soviet space hardware and systems was undertaken. It was concluded that a sizable proportion of the equipment then under development for the Moon effort could be reconfigured in relatively short order to accommodate the proposed space station requirement. By the end of the year, this had been formalized for the Korolev bureau to review, and by 1970, modification of various *Soyuz* components and the construction of what already was being called the *Salyut* ("Salute"; it was in fact a salute to Yuri Gagarin) space station's basic components was underway.

In general, the first-generation *Salyut* design was a simple arrangement of four short cylinders attached end to end with each having a different diameter and length. Three of the cylinders were designed to sustain livable, shirtsleeve environments for the cosmonauts, and the fourth contained the station's propulsion system.

(left) *The* Salyut *7 space station as it serenely orbits the Earth. To date, the Soviets have logged more than 20 man-years in space.*

EXHIBIT ARTIFACT: A 1:25 scale model of the Proton *booster. This launch vehicle has become one of the workhorses of the Soviet space program. Capable of placing 44,100 lb [20,000 kg] into low Earth orbit, it first entered operational service during 1965. Through 1989, it had 153 launch successes out of 178 attempts.*

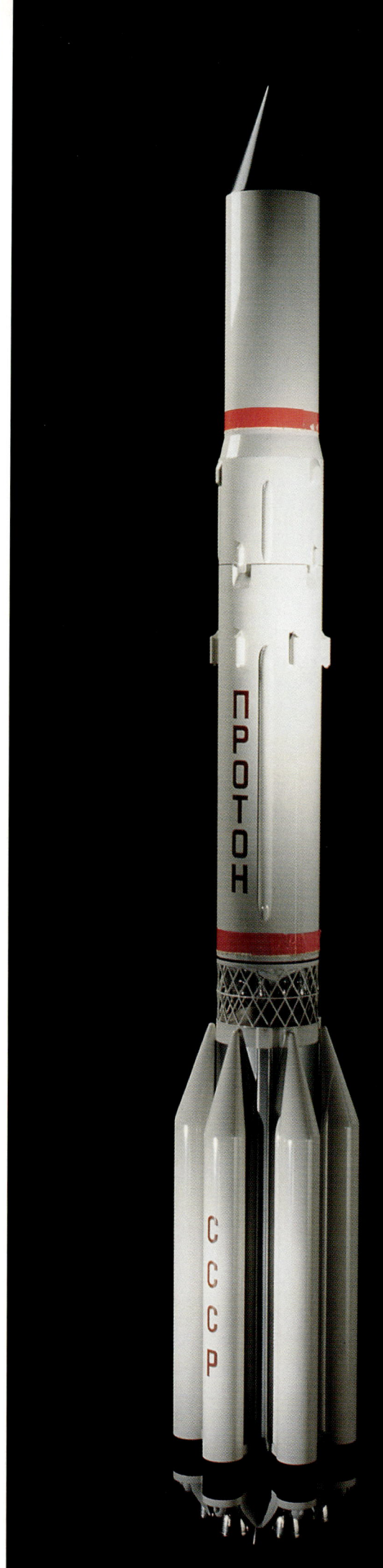

Proton *rocket*
1:25 scale model
Height: 145 ft 4 in [44.3 m]
SL-13 w/o payload
Weight: 1,486,170 lb
[674,000 kg] at launch
First launched: July 16, 1965

Integral with the *Salyut* design was the still viable *Soyuz* spacecraft which, as it turned out, was easily reconfigured from its original Moon mission to that of *Salyut* ferry. The biggest change to *Soyuz* was the installation of an internal crew transfer hatch system which allowed cosmonauts to crawl from the *Soyuz* into the *Salyut* without having to exit the interior of either spacecraft.

The transfer system consisted of a circular airtight door assembly that was hinged to fold into *Soyuz* after contact and docking with *Salyut*. Centrally mounted in the door was a probe which served as an alignment device during the docking process. This probe locked into a receptacle in the *Salyut* hatch and, via a screw-like action, pulled *Soyuz* into the proper docking position.

By the time of its change in mission objective, *Soyuz* was identified as the vehicle for transferring cosmonauts, research equipment, and supplies back and forth between the Earth and the forthcoming space station. Accordingly, it was reconfigured to dock at the front end of *Salyut* where the transfer compartment with associated airlocks was located.

Salyut, because of its innate adaptability, was capable of sustaining a large number of experiments. Accordingly, when first launched into orbit, more than 1,300 separate instruments were part of its payload. Included was an *Orion* 1 telescope for astrophysical observation, an *Anna* 3 gamma radiation telescope, an FEK-7 photoemulsion camera, micrometeorite sensors, an *Oasis* 1 hydroponic farm, physiological research tools, exercise items such as the *Penguin* suit, and a considerable quantity of biological research equipment.

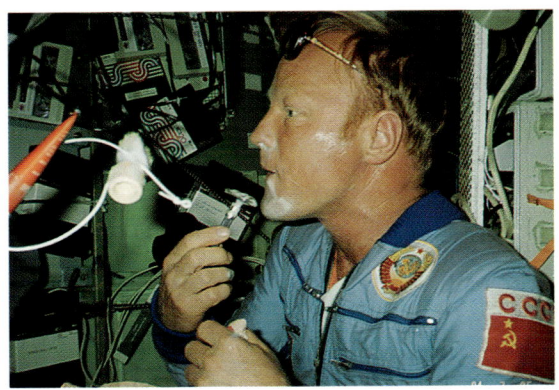

On April 19, 1971, some seven days after the tenth anniversary of Yuri Gagarin's historic first flight into space, a *Proton* booster successfully rocketed the world's first real space station, *Salyut* 1, into orbit. As planned, it was unmanned, waiting for the forthcoming arrival of the first *Soyuz* ferry.

A special cadre of nine cosmonauts by now had been picked and trained at Star City specifically for the *Salyut* missions. *Soyuz* 10, the first ferry to deliver cosmonauts to *Salyut* 1, was launched on April 23, 1971, but following docking on April 24, was unable to access the space station's interior. It is presumed that mechanical difficulties precluded successful completion of the mission. Regardless, it ended prematurely

(right) *In some respects at least, life aboard a space station is exactly like life on Earth.*

EXHIBIT ARTIFACT: Cosmonaut spacesuits. The Sokol suit on the left is that of cosmonaut Valery Kubasov from his Soyuz 19 (ASTP) mission. The Yastreb suit on the right is an earlier design. These are considered full-pressure suits as all parts of the body, including the hands and feet, are contained in an atmosphere-controlled environment.

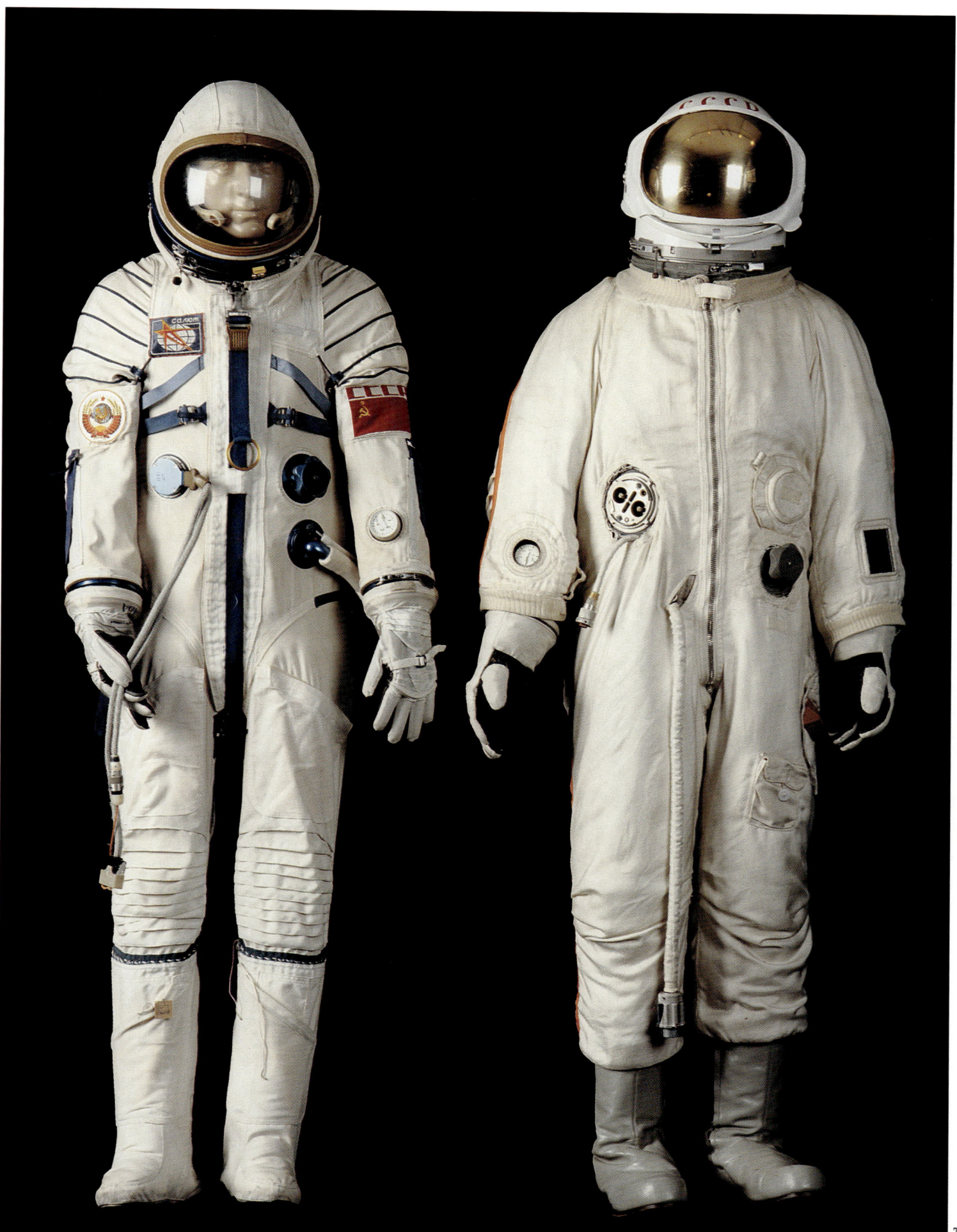

and the three cosmonauts onboard, Nikolai Rukavishnikov, V. Vladimir Shatalov, and Alexei Yeliseyev, were forced to return to Earth.

The first successful delivery of cosmonauts to *Salyut* 1 finally took place on June 7, the day after cosmonauts Georgi Dobrovolsky, Vladislav Volkov, and Viktor Patsayev were launched into orbit aboard *Soyuz* 11 mounted atop a *Soyuz* booster. The docking procedure this time went as planned and shortly afterwards, Patsayev became the first person ever to enter an orbiting space station from an Earth-to-space ferry.

The cosmonauts now spent a considerable period of time activating the new station's systems and undertaking a variety of experiments. Some three weeks eventually were spent aboard *Salyut*, with the return to Earth initiated on June 29.

Unfortunately, though the re-entry and module separation procedures went normally, by the time the recovery team located the *Soyuz* 11 capsule on the ground and opened its hatch, the three cosmonauts were dead. It later was concluded that a valve had opened inadvertently during the module separation event, and before emergency procedures could be introduced to correct the failure, all three cosmonauts had died from pulmonary embolism.

Partly in response to this failure and partly in response to the limitations of *Salyut's* onboard propellant capacity, on October 11, 1971, the station was de-orbited and allowed to re-enter the atmosphere over the Pacific Ocean. During re-entry it broke up and was destroyed.

During the year following the loss of the *Soyuz* 11 crew, design changes and improvements were introduced into the basic ferry configuration and work forged ahead on several different *Salyut* space stations. The new *Soyuz* design was tested during mid-1972 as *Kosmos* 496, and *Salyut* 2 followed nearly a year later, on April 3, 1973.

This space station, apparently with military systems onboard, proved unsuccessful and on April 14, according to some reports, began to break up in orbit. No manned *Soyuz* missions to *Salyut* 2 were attempted as a result, and little is known of its history or mission objectives.

However, with the upcoming launch of the first U.S. space station, *Skylab* 1, an attempt was made to get a second Soviet space station in orbit first under the guise of *Kosmos* 557. Strictly a propaganda move, it failed 11 days after launch and both *Salyut* 2 and *Kosmos* 557 were destroyed while re-entering the Earth's atmosphere in late May, 1973.

(above) *A* Proton *booster at launch. First stage propulsion consists of six RD-253 rocket engines and their side-module-type propellant tank assemblies attached to a cylindrical, central oxidizer tank. Introduced in 1965, the complete vehicle was not revealed to the West until the Vega missions of December, 1984.*

(middle) *Space station accommodations, including those of* Salyut *and the latest,* Mir, *are not overly capacious. Module cabins tend to be designed for multiple functions and virtually every square inch of wall space is utilized for equipment or storage.*

EXHIBIT ARTIFACTS: These lightweight jumpsuits were worn by the following cosmonauts: (left to right) Alexei Leonov, Soyuz 19; Boris Volynov, Soyuz 21–Salyut 5; Vladimir Dzhanibekov, Soyuz 27–Salyut 6.

(right) The Splav 02 *space furnace utilized on* Salyut 7 *space station, and also aboard* Photon, *was used primarily as a smelting kiln to study crystallization processes, production of metal alloys with exceptional uniformity, production of crystals from vapor and/or liquids, and numerous other metallurgical experiments in a weightless environment.*

EXHIBIT ARTIFACT: The Chibis *(left) and the* Penguin *physical conditioning suits. The* Chibis *suit is used by cosmonauts towards the end of a long-duration mission in order to return blood flow to the lower extremities to normal, Earth-like conditions. The Penguin suit is specially tensioned with elastic panels to help cosmonauts combat muscle deterioration during long duration space missions.*

On June 24, 1974, a replacement military space station, *Salyut* 3 was propelled into orbit after being launched from Baikonur Cosmodrome aboard a *Proton* booster. *Soyuz* 14 followed it into orbit on July 3 with Pavel Popovich and Yuri Artyukhin as crew. On July 4, the *Soyuz* docked with *Salyut* 3 and over the following two weeks, its cosmonauts conducted miscellaneous experiments and undertook a number of military missions. On July 19, *Soyuz* 14 returned safely to Earth, thus completing the first successful Soviet space station mission.

Like *Salyut* 2, *Salyut* 3 proved to have a relatively short life. On January 24, 1975, it was purposefully de-orbited and destroyed. *Salyut* 4, in the meantime, had been launched on December 26, 1974, and was declared another test laboratory for determining the viability of the space station's basic design parameters and systems, and for scientific experimentation in orbit. On January 11, 1975, *Soyuz* 17 was boosted into orbit aboard a *Soyuz* rocket and on January 12, docked to off-load its two cosmonauts. Alexei Gubarev and Georgi Grechko eventually spent 29 days aboard the space station, departing on February 9 for a safe return to Earth. Their tenure was the longest Soviet space mission on record at the time.

A follow-on mission to *Salyut* 4, *Soyuz* 18 proved unsuccessful when its *Soyuz* booster failed at an altitude of almost 100 miles [161 km] during a launch attempt on April 5.

Fortunately, both cosmonauts were saved via the *Soyuz* spacecraft's main propulsion system. Another *Soyuz* 18 launch followed on May 24 and was considerably more successful. Cosmonauts Pyotr Klimuk and Vitaly Sevastyanov stayed in space for 63 days, setting yet another Soviet record, and were followed to *Salyut* 4 by the unmanned *Soyuz* 20. The latter was designed to test automated control and docking systems related to the forthcoming *Progress* cargo delivery spacecraft, and also to test a proposed unmanned emergency rescue system. *Salyut* 4, like its predecessors, finally was de-orbited and destroyed on February 3, 1977, having been the most successful Soviet space station yet.

Salyut 5 proved the last of the first-generation Soviet space stations. Launched on June 22, 1976, as a military observation and test facility, it first was visited by *Soyuz* 21 on July 7 and remained occupied until August 24. On October 14, *Soyuz* 23 headed skyward aboard a *Soyuz* booster but its crew was destined never to board. Docking system difficulties terminated the mission on October 16 and it wasn't until the launch of *Soyuz* 24 on February 7, 1977, that a new crew was committed. The following day, cosmonauts Viktor Gorbatko and Yuri Glazkov

(below) *The manned space mission control center at Kaliningrad. A large projected video map display is surrounded by several slightly smaller textual information displays. Technicians man the conventional television monitors which provide specific data input.*

boarded the space station and reinstated the various research and observation programs that had been prematurely terminated by its previous occupants. On February 25, the original mission objectives were completed and *Soyuz* 24 undocked for its return to Earth.

Salyut 5 continued to provide research data for the following six months via remote systems and activities. However, on August 8, 1977, nearly 14 months after its launch from Baikonur Cosmodrome and after having hosted four cosmonauts and two ferry spacecraft, it was purposefully de-orbited. Following atmospheric re-entry, it broke up over the Pacific Ocean and was destroyed.

This event marked the end of the first-generation *Salyut* space stations. New systems, an improved design criteria, and the experience base that had been derived from the first five *Salyuts* and their ground-based analogs resulted in a new *Salyut*. This spacecraft, though superficially resembling its predecessors, was in fact an almost totally revamped design.

With the advent of the second-generation *Salyuts*, the Soviets introduced a new system for ferrying fuel, supplies, test equipment, and cosmonauts to their space stations. Because nearly 20 tons of supplies were required to support a *Salyut* in space over a two-year period, such a system was mandatory for successful space station maintenance.

The *Progress* ("Progress") ferry by 1973 had entered initial stages of development. Based on the dependable and thoroughly understood *Soyuz*, it differed primarily in being unmanned, remotely controlled, and optimized for the transportation of supplies (rather than cosmonauts).

Complementing *Progress* was a second new ferry-type spacecraft referred to in the West as *Heavy Kosmos* and having a re-entry module similar in shape to the U.S. *Gemini* manned space capsule. Unlike *Gemini*, however, it would have been reusable. *Kosmos* 929, 1267, and 1443 were the only missions eventually flown with this spacecraft, which also had a large working compartment suitable for use as an extra work space when attached to a *Salyut* space station.

A third new ferry also was introduced during this period and referred to as *Soyuz* T. Until its advent, all *Soyuz* missions conducted after *Soyuz* 11 had carried only two cosmonauts. The new ferry could accommodate three cosmonauts per flight and was capable of sustaining them for up to four days (the earlier *Soyuz* carried only enough oxygen and electrical power for two days in space). Additionally, the new ferry was more automated than its predecessors. It also had a more versatile propulsion system that included the ability to utilize propellants from the main engine to power the in-space maneuvering thrusters.

(below) *A* Soyuz *booster on its railroad transport being moved to the launch pad at Baikonur.*

A *Heavy Kosmos* was the first of the new ferries to be placed in service when it was boosted into orbit on July 17, 1977. *Progress* 1 followed on January 20, 1978, and in turn was followed by the first *Soyuz* T on December 16, 1979 (however, *Soyuz* T had been tested as a *Kosmos* project as early as November of 1976). The latter two systems have been upgraded and remain in service as of this writing.

Launched on September 29, 1977, aboard a *Proton* booster, the first of two second-generation Soviet space stations, *Salyut* 6, was provided with the mechanical and physiological systems required of longer and considerably more complex space missions. Though retaining the basic shape of *Salyut* 4, the new station differed in being equipped with both front and rear docking modules— these permitting two ferry spacecraft to service it at any one time. The basic configuration, according to Soviet literature, consisted of a forward transfer compartment, the first work compartment, a connecting frustrum, a second work compartment, and a service compartment. Each compartment was fully pressurized.

Salyut 6 was placed in orbit as a civilian platform and accordingly was equipped for non-military research missions. It was the first Soviet space station to have a fully-integrated "Orientation And Motion Control System" to automatically orient the station as required for various experiments. This same system also permitted the correction of orbit anomalies to facilitate rendezvous and docking maneuvers.

Onboard experiments carried by *Salyut* 6 were vast in number, but among the more important were the MKF-6M Earth resources camera for photographing the Earth in six spectral bands at once; the KATE-140 wide-angle, stereographic/topographical camera for color map making; and the BST-1M telescope for recording atmospheric data in the infrared, ultra-violet, and sub-millimeter spectral ranges.

Weighing 41,675 lb [18,900 kg] at launch, *Salyut* 6 had been designed from the start to accommodate lengthy stays in space. Accordingly, two manned mission types were planned for the new space station, these being divided between EO (*Ekspeditsya osnovnoi*/"Principal Expedition") and EP (*Ekspeditsya poseshchenya*/"Visiting Expedition") crews. The EO crews later were divided between EO-1 with a scheduled stay of 90 days; EO-2 with a scheduled stay of 120 to 140 days; and EO-3 with a scheduled stay of 180 days.

EP crew stays were varied and were dictated by the two-and-a-half month viability of the *Soyuz* ferry once it was docked to the

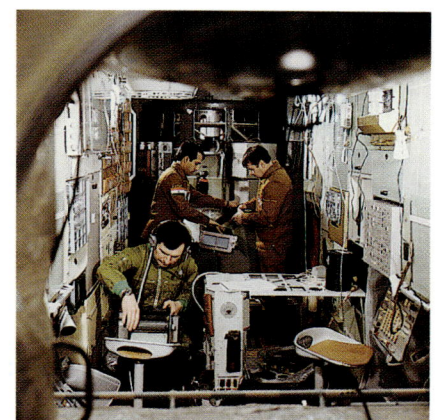

(above) *Life inside a space station is not boring, but it can be monotonous during long-duration missions. Daily routines are scheduled well in advance of a mission and include rest and relaxation activities such as television communication with relatives and friends. Ferry missions often bring gifts from wives and children as well as essential cargo.*

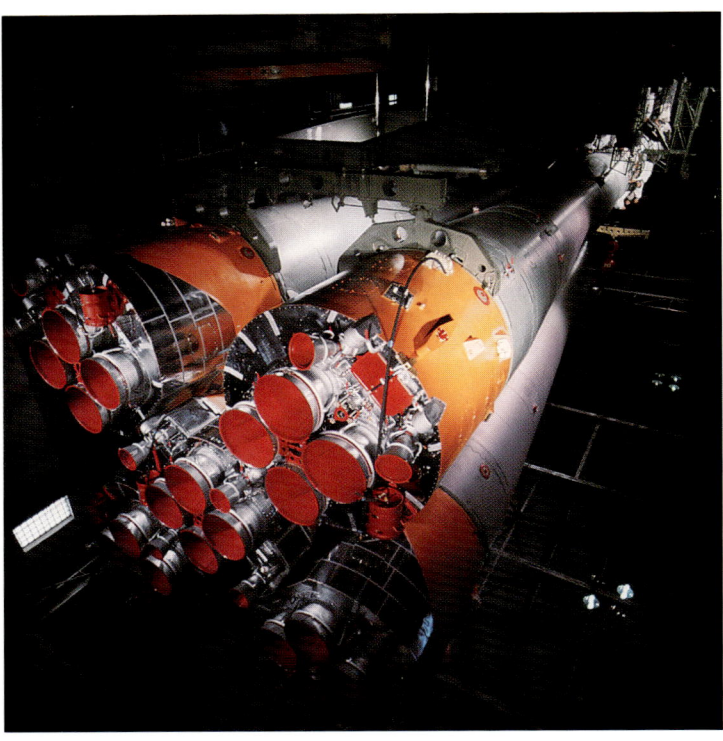

(above) *The gantries surrounding this* Soyuz *rocket provide crew and launch pad personnel access to the vehicle. Pre-dawn activity at Baikonur Cosmodrome requires powerful artificial lighting capability for maintenance proceedings.*

(middle) *The clustered engine configuration visible here is used in the* Vostok, Molniya *and* Soyuz *boosters.*

space station. Accordingly, a fresh *Soyuz* was required to replace the docked *Soyuz* approximately every ten weeks if the latter had not been used by the predetermined "unsafe" date. Thus, most EP-type missions were of short duration and designed essentially to exchange the old *Soyuz* ferry for the new.

As part of the *Salyut* 6 itinerary, the long-standing Soviet *Interkosmos* agreement calling for cooperative space exploration projects involving Bulgaria, Czechoslovakia, Cuba, the (former) German Democratic Republic, Hungary, Mongolia, Poland, Romania, and (later) Vietnam was redefined to include the use of Soviet-trained cosmonauts from those countries. These scientists and engineers, of which there would be many, would occupy what originally had been scheduled to be an empty seat in the *Soyuz*-to-*Salyut*-and-back resupply and EP missions.

Following *Salyut* 6's successful insertion into orbit, the first *Soyuz* ferry flight to carry a crew to the new space station, *Soyuz* 25, was launched on October 9. Unfortunately, docking difficulties prevented the two-cosmonaut crew, Vladimir Kovalenok and Valery Ryumin, from boarding, and it wasn't until the launch of *Soyuz* 26 with Yuri Romanenko and Georgi Grechko onboard on December 10, 1977, that *Salyut* 6 at last was manned.

The docking problems of *Soyuz* 25 remained a concern, however, as there was some possibility that *Salyut* 6's front docking module was where the difficulties lay (*Soyuz* 26 had docked at the rear and thus could not verify if a problem existed). An interim *Soyuz* delivery flight, *Soyuz* 27, manned by Vladimir Dzhanibekov and Oleg Makarov later verified the integrity of the front unit, and the mission continued without further difficulties.

On January 20, 1978, the first *Progress* unmanned cargo craft was launched carrying food, oxygen, water, film, and propellant for the space station's rocket engine. Following docking on January 22, all conventional supplies were quickly off-loaded by the *Salyut* 6 crew. Off-loading of propellants took considerably longer and the job was not completed until February 1. On February 6, after being loaded with *Salyut's* waste material, *Progress* was undocked, and eventually de-orbited to be destroyed during the re-entry process over the Pacific Ocean.

Events aboard *Salyut* 6 now became essentially routine with the various crews setting endurance records on an almost daily basis and with experiments and photography filling up their many monotonous hours. What follows is a list of the various missions that took place during the course of *Salyut* 6's lengthy EO-1, EO-2, and EO-3 space tenure after the arrival and departure of *Progress* 1.

EXHIBIT ARTIFACTS: The tools used by cosmonauts for maintenance, repair, and assembly work aboard the various Soviet space stations are considerably more complex and versatile than their conventional Earth-bound stablemates.

(below) *Because it must be consumed in a weightless environment and also because of limited storage space, most of the food consumed aboard the various Soviet spacecraft and space stations are highly processed and formulated for an extended shelf life. Tube-like containers are commonplace and anything that needs to be heated before consumption is designed to fit in a small electric stove.*

EXHIBIT ARTIFACT: All world aerospace records are sanctioned by the Federation Aeronautique International (FAI). When Anatoly Berezovoi and Valentin Lebedev completed their 211 day long-duration mission aboard Salyut 7, it was, at the time, the longest anyone had stayed in space. The FAI later accepted the two cosmonauts' record claim. This leather-bound volume contains official FAI confirmation documents.

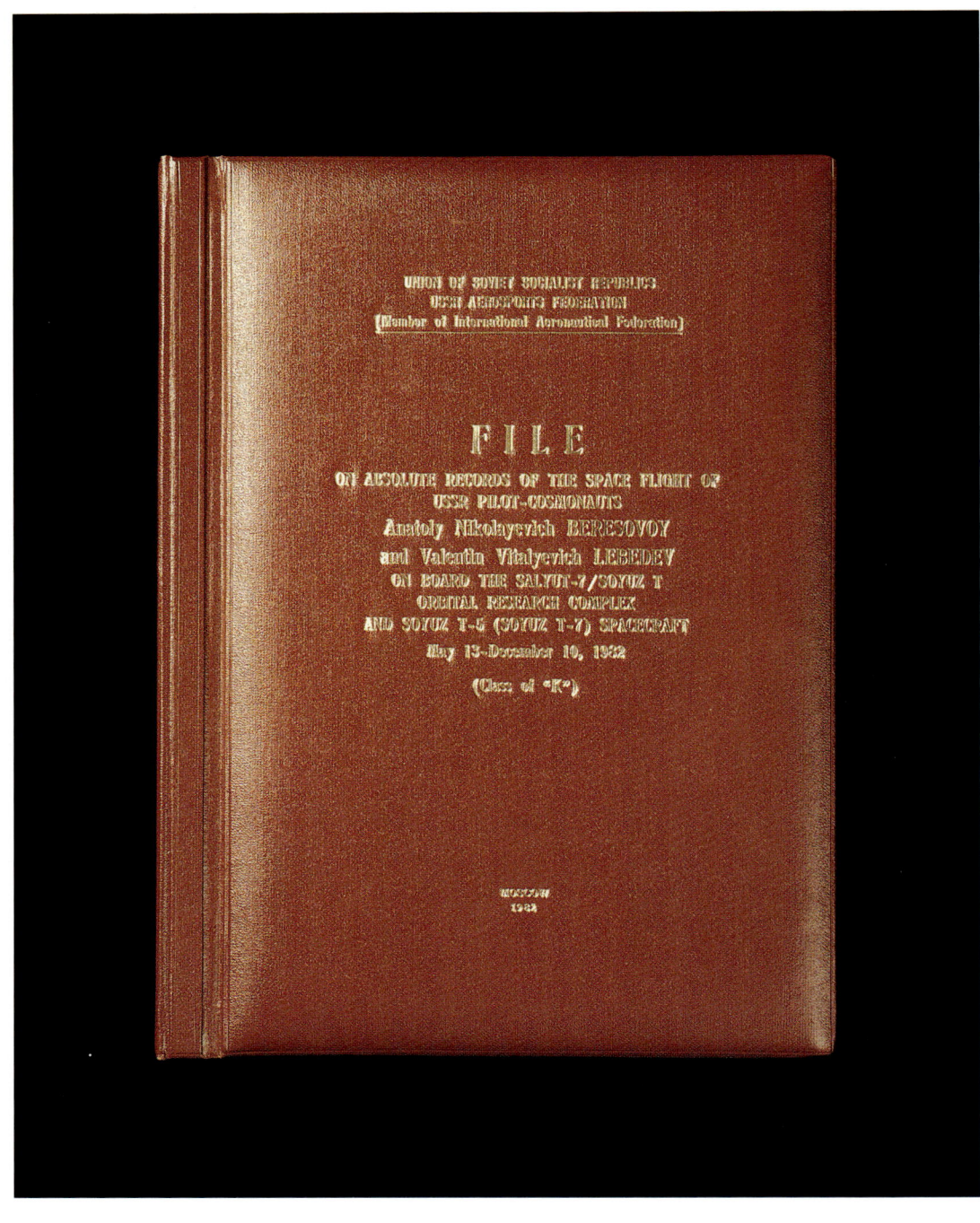

· On March 3, 1978, *Soyuz* 28 arrived carrying the first international *(Interkosmos)* cosmonaut, Vladimir Remek from Czechoslovakia.

· *Soyuz* 29 started EO-2 on June 16 when it docked with *Salyut* 6.

· *Soyuz* 30 heralded the arrival of the first Polish cosmonaut, Miroslav Hermaszewski when it docked on June 28.

· *Progress* 2 made its remotely-controlled appearance on July 9 in order to deliver food, water, miscellaneous supplies, and propellant.

· *Soyuz* 31 successfully completed the first in-space transfer of a *Soyuz* from one *Salyut* 6 docking port to another on September 7.

· On October 6, *Progress* 4 arrived with supplies and propellant.

· On November 2, *Soyuz* 31 began the shuttle flight carrying the EO-2 crew back to Earth.

· On February 26, *Soyuz* 32, after docking with *Salyut* 6, initiated EO-3 which was scheduled to last some 175 days.

· On March 14, *Progress* 5 brought supplies and propellant.

· On April 11, *Soyuz* 33 became the first *Soyuz* to suffer from main propulsion system failure . . . and as a result was forced to return to Earth with its civilian commander and the first Bulgarian cosmonaut still onboard.

· On May 15, *Progress* 6 brought the usual supplies and propellant and made up somewhat for the failure of *Soyuz* 33.

· On June 8, *Soyuz* 34 brought miscellaneous supplies.

· On June 30, *Progress* 7 arrived carrying a folded KRT-10 radio telescope which was to be tested prior to the end of the EO-3 mission.

· On August 19, 1979, *Soyuz* 34 returned to Earth carrying EO-3's crew, Vladimir Lyakhov and Valery Ryumin after their having spent 175 days, 36 minutes in space.

Though originally expected to last only two years, *Salyut* 6 at the end of EO-3 was re-evaluated and determined still to have considerable life left. As a result, the first of the *Soyuz* T missions (unmanned) was rocketed into orbit on December 16, 1979. It was to dock with *Salyut* 6 using remote control and then use its propulsion system to re-invigorate the space station's rapidly decaying orbit by giving it a propulsive boost. As *Salyut* 6's own propulsion system had ceased to function, this was the only way to save it from re-entry and destruction. Such periodic boosts were mandatory in order to keep the station in space.

After a follow-up orbital boost by an unmanned *Progress* spacecraft, *Salyut* 6 was remanned on April 10, 1980, by Leonid

«Ракета «Союз» оторвалась от Земли, стали нарастать перегрузки, и я почувствовал себя совсем другим человеком. Мне стало казаться, что нет силы, которая может вернуть меня обратно.»

Валерий Рюмин

"The *Soyuz* rocket tore itself from the Earth's clutches, the force of the G's began to grow, and I felt myself change into another person. It seemed to me that there was no force in the universe that could ever return me to Earth."

Valery Ryumin

(above) *The effects of weightlessness in space, while appearing humorous in this picture of a cosmonaut and a floating watermelon, are insidious and can have a disabling effect on the human body if not checked by rigorous exercise routines.*

EXHIBIT ARTIFACT: Soyuz-Salyut *scale model. Through missions to the Salyut space stations, the Soviets have learned a great deal about the long-term effects of weightlessness on the human body.*

Soyuz-Salyut *complex*
1:100 scale model
Launched: April 19, 1971 - Salyut 1
Launch vehicle: Proton
Weight: 72,148 lb [32,720 kg] as shown

Popov and Valery Ryumin who arrived aboard *Soyuz* 35 and quickly brought it back on line. During the following year, it continued to successfully perform its space station function and was visited by a number of different Soviet and international cosmonauts. Finally, on May 26, 1981, it was abandoned for the last time when the last remaining cosmonauts, Vladimir Kovalenok and Viktor Savinykh, entered a *Soyuz* T ferry and departed for their return mission to Earth. A year later, on July 29, 1982, after performing remotely-controlled experiments in association with several different *Kosmos* satellites, *Salyut* 6 was de-orbited and destroyed during re-entry over the Pacific Ocean.

Salyut 6 proved to be a milestone in the history of early space station development due to its longevity and to its mission accomplishments. It hosted no less than 35 cosmonauts and continued, like its first-generation predecessors, to explore the boundaries of long-duration space missions.

Salyut 7, though initially anticipated to be a third-generation space station, was in fact virtually identical to *Salyut* 6. *Salyut* 7 was launched by a *Proton* booster on April 19, 1982, just over 21 years after Gagarin's historic first space flight. Eventually it would last some nine years in space, be manned by a large number of Soviet and non-Soviet cosmonauts, witness the greatest number of space walks on one space station mission ever performed, serve as a launch pad for the *Iskra* 2 and 3 amateur radio sub-satellites, and later be abandoned not to an ignominious end over the Pacific Ocean, but to a storage orbit at an altitude of 295 miles [475 km]. There it would stay until February 6, 1991, when solar activity effects finally caused its orbit to decay to the point where an uncontrolled re-entry over South America was unavoidable.

During the course of *Salyut* 7's life, it was manned by a number of long-duration crews, including that of a *Soyuz* T ferry launched to the space station on May 13, 1982, and consisting of Anatoly Berezovoi and Valentin Lebedev. These two cosmonauts eventually would spend a total of 211 days, 8 hours, 5 minutes in space, not returning to Earth until December 10. Lebedev's experiences would later be chronicled in his insightful *Diary Of A Cosmonaut: 211 Days In Space*.

Operations and experiments performed aboard *Salyut* 6 and *Salyut* 7, coupled with the experience base generated by their numerous cosmonaut crews, proved of great benefit when the Soviet space team sat down to begin design of their third generation space station, *Mir* ("Peace," "World," or "Commune") during the mid-1970s. Created as a modular system offering considerable versatility in terms of assembly, structure, and utilization, it consists of a core station resembling the shell of the second-generation *Salyuts*. In turn, this consists of two intermediate length cylinders and a connecting frustrum. These cylinders, which make up the working compartment, are in turn attached by the frustrum to a ball-shaped pressurized transfer adapter. On the other end, attached to the working compartment, is an unpressurized service compartment. Docking ports are available at either end of the longitudinal core system, and also at four axial points of the multiple docking adaptor.

Like its predecessors, *Mir* is autonomous from an electrical power standpoint and is equipped with a large solar panel assembly mounted on two massive, extendible module panels. Attitude control is accommodated by 32 thrusters, and orbital maintenance is accommodated by two 661 lb [300 kg] thrust rocket engines. When the complete *Mir* complex is assembled in space, the weight is expected to total 140 tons [127,000 kg].

Like the second-generation *Salyut* stations, *Mir* has benefited from the development of several new ferry spacecraft. The first of these is the *Soyuz* TM ("Modified") which, though resembling earlier *Soyuz* spacecraft, is equipped with an array of greatly improved systems and structural components. It can carry several times the return payload of its predecessors.

The unmanned *Progress* ferries also apparently have undergone some changes and have increased payload and propellant capacities. A third spacecraft of totally new configuration has entered the space fleet under the name *Kvant* ("Quantum"). It is essentially an astrophysics module and is designed to dock permanently with *Mir*, consequently expanding its size and research capability.

The first, and to date only, *Mir* space station was rocketed into space aboard a *Proton* (SL-13) booster on February 19, 1986. On March 13, the first two-cosmonaut crew, Leonid Kizim and Vladimir Solovov, seated in their *Soyuz* T-15 ferry, were launched by a *Soyuz* rocket into space to rendezvous for the first time with *Mir*. Docking at the space station's front module took place on March 15 and within an hour of arrival, the crew had entered and demothballing procedures had been initiated.

The initial *Mir* stay proved relatively short-lived, for on May 3 it was announced that the two-cosmonaut crew would be leaving the space station to visit the nearby *Salyut* 7/*Kosmos* 1686 complex. On May 5, *Mir* was abandoned and on May 6, *Salyut* 7 was revisited. This precedent-setting first excursion from one space station to another was to mothball *Salyut* 7 in preparation for its forthcoming movement into a permanent parking orbit.

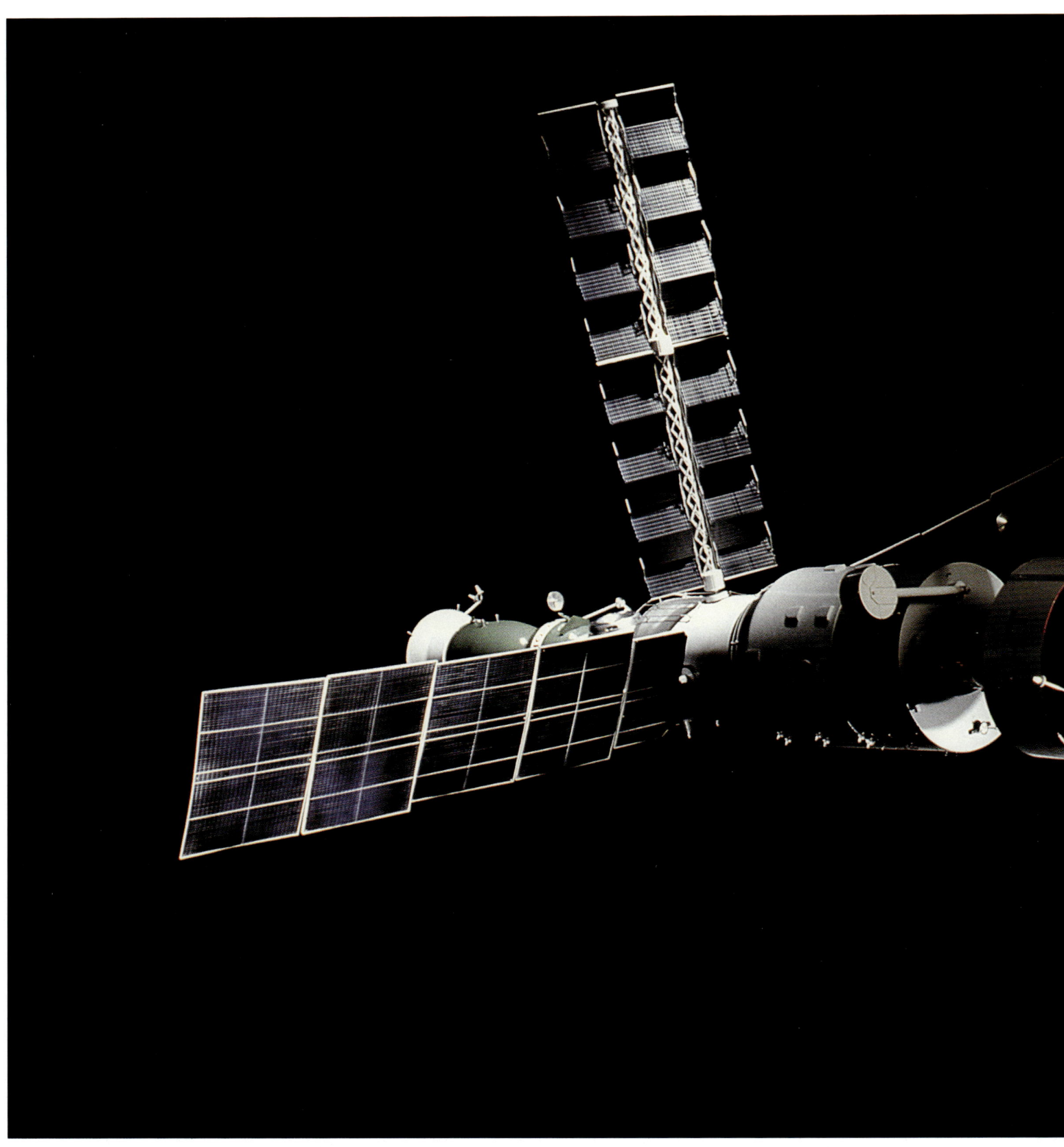

EXHIBIT ARTIFACT: Mir *has become the most successful of the several Soviet space stations launched to date. This scale model shows the complex as it appeared in 1987. Since then, the addition of the* Kvant 2 *and* Kristall *modules has considerably expanded the volume of the station.*

Mir *orbital station*
1:8 scale model
Launched: February 19, 1986 (main module)
Launch vehicle: Proton
Weight: 100,769 lb [45,700 kg] as shown

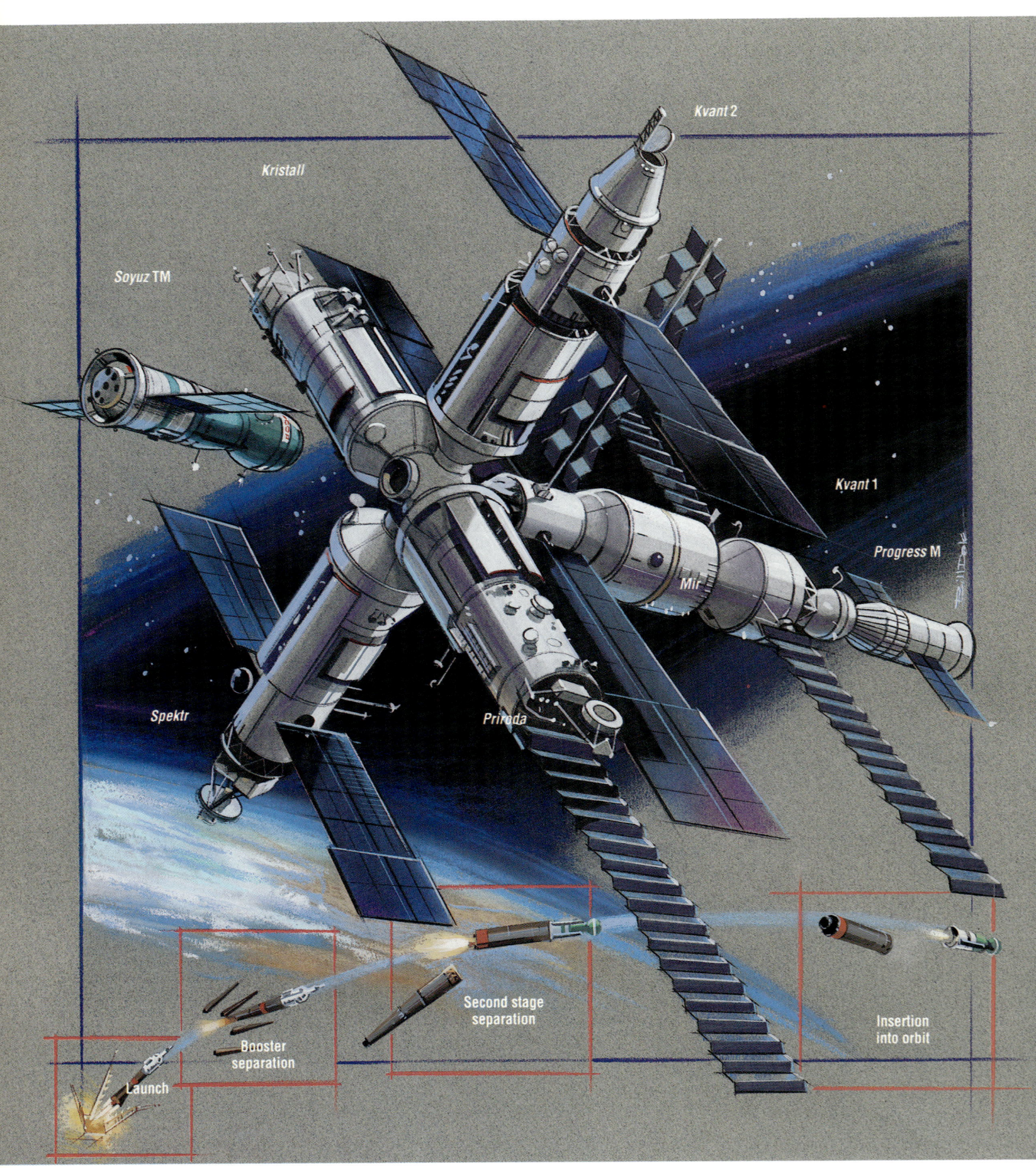

(above) The Mir *space station, as it will look in 1992, consists of two different* Kvant *and various other modules interfaced using an axial docking port and other interfacing assemblies. It is supplied with fresh crew members on a regular basis using* Soyuz TM *space ferries. More mundane supplies such as food and propellant arrive aboard unmanned* Progress M *ferries.*

In the interim, work on *Mir* continued, albeit remotely, via the launch of an unmanned *Soyuz* TM spacecraft. On June 26, the two original cosmonauts returned from their *Salyut* 7 mission and once again set up shop inside *Mir*. This visit, too, proved short-lived, and on July 16, they returned to Earth.

Throughout the rest of 1986, *Mir* circled the globe in a mothballed and unoccupied state. The main reason for this was due to the delay in delivery of the *Kvant* module. During January of 1987 an unmanned *Progress* ferry was sent to the station to deliver propellant and supplies. Docking automatically, it then awaited the arrival of two cosmonauts, Yuri Romanenko and Alexander Laveikin, launched on February 5. Docking of the latter in their *Soyuz* TM-2 ferry took place on February 7, and following their entry into the space station, they began the difficult task of manually off-loading the supplies that were waiting in the *Progress* spacecraft.

On March 31, following several weeks of routine activity aboard *Mir*, a *Proton* booster rocketed the first *Kvant* module into orbit. *Kvant* was the first *Mir* expansion module and weighed a total of 22.7 tons [20,590 kg] with its propulsion unit. It carried test instrumentation weighing approximately 3,300 lb [1,500 kg]. Designed to serve as an integral *Mir* assembly following docking, it ran into difficulty and was not safely docked until the two *Mir* cosmonauts, Romanenko and Laveikin, took a space walk and manually removed blockage in the form of a plastic bag stuck between the *Kvant* attachment probe and the *Mir* receiving drogue.

On July 22, 1987, *Soyuz* TM-3 was launched, carrying cosmonauts Alexander Victorenko, Alexander Alexandrov, and the first international guest, Mohammed Faris of Syria. These men joined Romanenko and Laveikin aboard *Mir* and stayed until the arrival of *Soyuz* TM-4 which brought cosmonauts, Vladimir Titov, Musa Manarov, and Anatoly Levchenko on December 23.

On July 29, Victorenko, Faris, and the veteran Romanenko returned to Earth. Romanenko by now had logged some five months in space and concerns about his health had been expressed due to a spate of heart irregularities. These later proved unfounded; post-flight examinations revealed nothing abnormal and he subsequently was returned to flightworthy status.

The following year, cosmonauts Vladimir Titov and Musa Manarov completed respective stays of a year in space.

(right) *Because of the deleterious effects of weightlessness on the human body, it is imperative that a healthy exercise regimen be maintained. Accordingly, all Soviet space stations have been equipped with an extensive array of exercise equipment to accommodate human muscle and cardiovascular requirements.*

«Мой сын Виталик ласкается, целует меня, — чувствует, что отец улетает надолго. Перед уходом из дома мы сели за стол и по традиции оставили хлеб, соль и воду...когда отъезжали, я посмотрел назад и увидел, что мама вытирает слезы. Я помахал ей рукой, но она меня не видела...»

Валентин Лебедев

"My son Vitalik snuggles next to me and kisses me; he realizes that his father is going away for a long flight. Before we left the house we sat down at the table and followed our tradition of leaving out bread, salt, and water... As we drove off, I looked back and saw that Mother was wiping away tears. I waved to her, but she did not see me..."

Valentin Lebedev

EXHIBIT ARTIFACT: This full-scale mock-up of a Soyuz return module is just one of three parts making up the total Soyuz package. Normally, in stacked condition, the orbital module is mounted on top, the return (or re-entry) module is in the middle, and the instrument module is on the bottom.

Soyuz return module
First launched: November 28, 1966
Launch vehicle: Soyuz
Weight: 6,178 lb [2,802 kg] - Soyuz 19

(right) Soyuz *launch and recovery.*

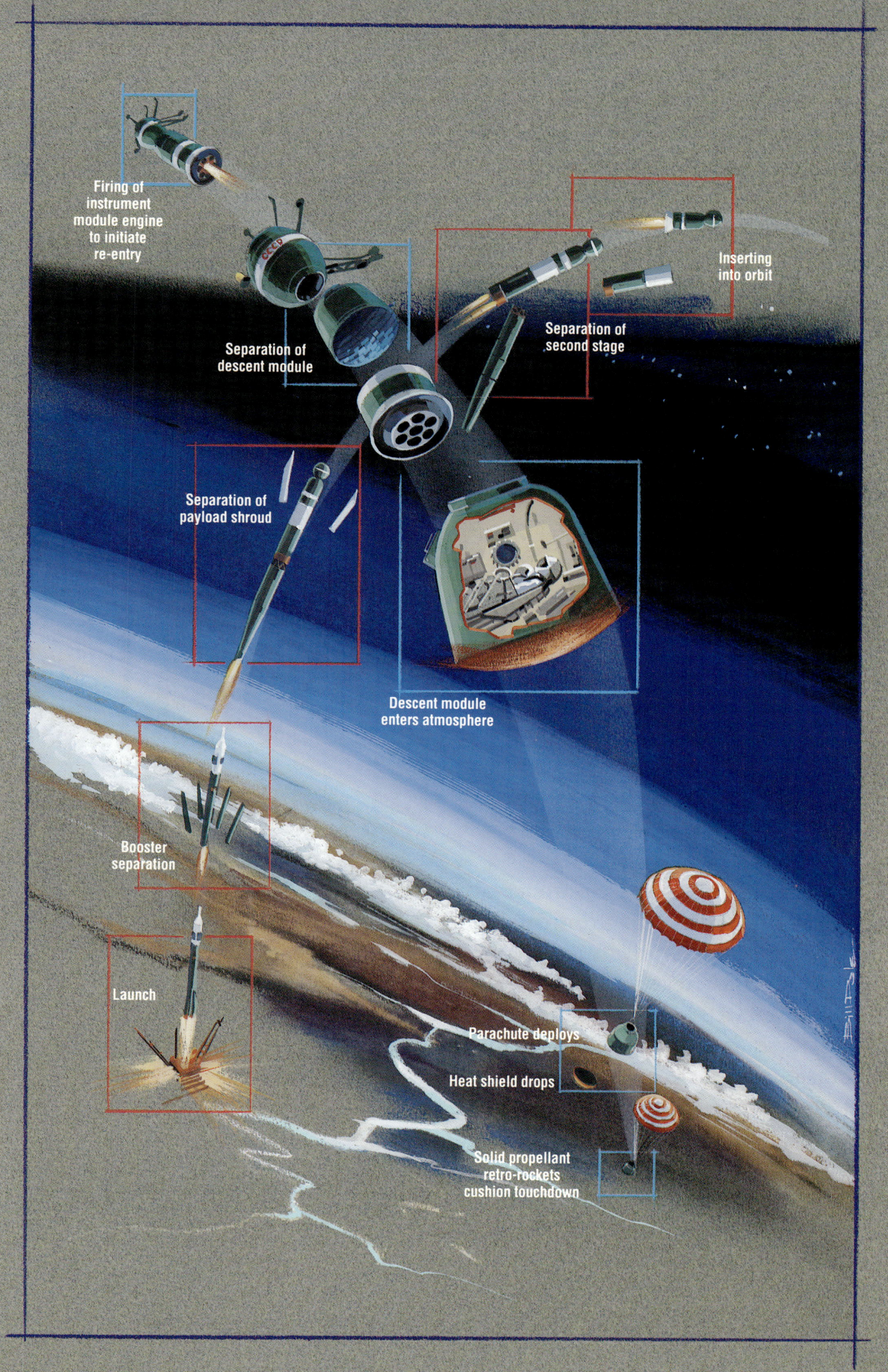

EXHIBIT ARTIFACT: A contemporary Orlan *spacesuit and* Ikarus *"space motorcycle". The* Ikarus *is a compressed air rocket propelled backpack-type manned maneuvering unit designed to give a cosmonaut autonomous mobility in a space environment. Though the* Ikarus *has been successfully tested in space, all EVAs have been tethered so far, due to the high risk involved.*

INTERNATIONAL COOPERATION IN SPACE

Intermingled with the successes being enjoyed by the *Salyut* space stations and their associated *Soyuz* ferry missions, peripheral work with various *Kosmos* flights and *Soyuz* flights that were only indirectly related to *Salyut* also were undertaken. Three *Soyuz* missions, *Soyuz* 16, 19, and 22, all were directly related to what had become known as the *Apollo-Soyuz* Test Project, or ASTP.

ASTP was an end product of the state of *detente* that had arisen between the Soviet Union and the U.S. during the mid-1960s. It was, in reality, a visible gesture of peace between the two superpowers, and thus in the public eye, a worthy goal.

Timing of the project proved propitious for both parties concerned. By the early 1970s, the U.S. space program was losing impetus and becoming mired in a state of public apathy; and similarly the Soviet space program was in need of something that would boost the lackluster image that had surfaced as a result of the *Soyuz* 11 disaster and the loss of the three cosmonauts.

Following many meetings between NASA personnel and representatives from the Soviet Union's Academy of Sciences, the basic objectives and technical details of a joint *Apollo-Soyuz* mission were worked out. On May 24, 1972, an agreement to consummate the flight during 1975 was signed by President Richard Nixon and Premier Alexei Kosygin. Over the following three years, considerable effort and money were expended by both countries to develop the critical equipment required to make the mission successful.

(left) Mir, *essentially as it looks today. Several additional large modules are scheduled to be added by the time the space station is completed near mid-decade. Total weight at that time will be in excess of 140 tons [127,000 kg].*

The single most important item was the docking module. This complex device was the interface that permitted the cabin atmospheres of the two decidedly different spacecraft to be mixed. Concurrently, it provided the compatible attachment mechanisms for the two mechanically-different docking units.

On July 15, 1975, Soviet cosmonauts Alexei Leonov and Valery Kubasov were rocketed into space in their special *Soyuz* 19 spacecraft mounted atop a *Soyuz* booster. Several hours later, *Apollo* 18, mounted atop a *Saturn* 1B booster, carried astronauts Thomas Stafford, Vance Brand, and Donald Slayton into orbit. Along with them went the complex docking module which, after orbit was achieved, was moved into position on top of the U.S. spacecraft.

On July 17, 1975, the historic docking took place when *Soyuz 19* and *Apollo* 18 made contact. Some three hours later, the *Soyuz* docking hatch was opened and within seconds, Stafford and Leonov were shaking hands. It was a noteworthy and highly publicized event, and the first step in the direction of what the world hoped would be many more cooperative space ventures.

The two spacecraft remained attached for nearly 48 hours. In the interim, the cosmonauts and astronauts spent considerable time together working on experiments and communicating ideas. A separation and second docking also was undertaken near the end of the visit, with a final separation taking place on July 19. On July 21, *Soyuz* 19 returned safely to Earth; it was followed on July 24 by *Apollo* 18.

ASTP was not the first nor the only cooperative space venture in which the Soviets had participated by 1975. In fact, their notable *Interkosmos* and *Intersputnik* projects had encompassed a virtual plethora of internationally sponsored manned and unmanned technological, biological, astronomical, meteorological, and planetary space studies, and eventually they would provide the entree to many other internationally flavored satellites and probes that continue to be launched to this very day.

Separate from, but integral with the international satellite projects undertaken by the Soviets are their numerous military and civilian navigation satellites. These spacecraft, optimized to provide extremely accurate navigational information, can be accessed from virtually any spot on the globe. *Tsikada* encompasses the civilian navigation system.

Integral with the *Tsikada* Soviet navigational satellite program is the *Nadezhda* satellite which is associated with the Soviet segment of the international search and rescue system (KOSPAS in the Soviet Union and SARSAT in the U.S.). Designed to determine the accurate location of ships and downed aircraft in distress, KOSPAS systems have been carried by five satellites and are capable of accurately relaying to ground rescue teams the position of any of the 350,000 emergency beacons currently in use. Coupled with similar systems carried by Western satellites, any activated rescue beacon can be accurately located and positioned anywhere in the world.

Perhaps most importantly, the Soviets became the first to share space stations when they made *Salyut* and *Mir* available to foreigners. Cosmonauts from Bulgaria, Cuba, Czechoslovakia, Egypt, Germany, Hungary, India, Iran, Japan, Mongolia, Poland, Romania, Syria, Vietnam, Britain, and numerous other countries since have been, and continue to be, invited to participate in space station activities and research.

All of these co-ventures have been and continue to be conducted with an eye toward peace and good will among nations. The resulting positive publicity continues to generate a high degree of support for the Soviet space program from non-Soviet countries. ●

(below) *There is not much elbowroom inside a space station.*

EXHIBIT ARTIFACT: A scale model of the Energiya *launch vehicle. Large central core is the main propellant tank and is equipped with four liquid-oxygen/liquid-hydrogen engines and their associated turbo-pump assemblies. The four surrounding strap-on liquid-fuel boosters each have a single liquid-oxygen/kerosene engine with four exhaust nozzles.*

Energiya *rocket*
1:50 scale model
Height: 196 ft 10 in [60 m]
Weight: 5,290,000 lb [2,400,000 kg]
First launched: May 15, 1987

«Мы летели над Америкой. Я ощутил холодок раннего осеннего утра, тишину и грусть мерзлой Земли. Я никогда не был в Америке, но увидел то, что много-много раз чувствовал на Родине. Первый снег, золотые листья осени — все одинаковое и родное для нас на всей Земле. Мы — дети ее.»

Александр Александров

"We were flying over America. I felt the chill of an early autumn morning, the silence and sadness of the frozen Earth. I had never been in America, but I saw something first—snow and the golden leaves of autumn—these are equally dear to all of us all over the entire Earth. And we are all her children."

Alexander Alexandrov

(above) *The historic first handshake of the* Apollo-Soyuz Test Project (ASTP) *was undertaken by Tom Stafford of the U.S. and Alexei Leonov of the Soviet Union on July 17, 1975.*

(middle) *Patches commemorating the successful* Apollo-Soyuz Test Project. *Outline of Soyuz is visible on lower left segment of Soviet patch (left), and outline of Apollo is visible on lower-left of U.S. patch (right).*

SOVIET SPACE PROGRAM TODAY AND TOMORROW

The apparent successes being enjoyed by the U.S. space shuttle program which was launched into orbit for the first time on April 12, 1981, led the Soviets to reassess their own space shuttle project which had been started as early as 1963, but had never progressed to a full scale hardware stage. The appeal of such spacecraft lay in their reusability and theoretically cheaper operating costs. For this reason, Soviet interest not only was renewed, but bureaucratically encouraged.

The Soviet Union, like several of its Western counterparts, had evaluated the attributes of manned, maneuverable, reusable spaceplanes for many years, but unlike the U.S., had elected to continue utilization of the more conventional capsule-type space systems dependent upon large boosters for insertion into orbit. This system, as time went on, was found to be prohibitively expensive. Not only were the boosters not reusable, but the actual spacecraft were good for only one mission as well. Exploratory work with *Heavy Kosmos* tested the idea of a reusable re-entry capsule, but three-unmanned tests during 1977, 1981, and 1983 provided mixed results.

Accordingly, Soviet interest in the shuttle concept continued to resurface and various studies were undertaken to explore the shuttle idea. Several of these, including at least one generated by the famous Mikoyan and Gurevich (MIG) fighter design bureau, resulted in the flight testing of a small-scale prototype. Difficulties with the MIG configuration prevented the maturation of strong Kremlin support, however, and by 1969 it had been abandoned.

Five years later, V. P. Glushko, one of the fathers of Soviet rocketry and an engineer whose pioneering propulsion system efforts could be traced back to the 1930s, proposed a more expansive space shuttle somewhat along the lines of the vehicle that then was being conceived in the U.S. Consequent to this, the death of Korolev during 1966 permitted Glushko to move into the Soviet space program seat of power. As the new "Chief Designer of Rocket-Cosmic Systems", he quickly replaced a

(above right) *With its modular engineering,* Energiya *may fly with as many as eight strap-on boosters in the future.*

(right) *The* Buran *space shuttle has been launched as an* Energiya *payload on only one occasion. This initial flight, which took place on November 15, 1988, was successful, but unmanned. The entire mission was remotely-controlled.*

Still further into the future, the Soviets are exploring the possibility of manned missions to Mars, with launch dates as early as 2005. In order to facilitate the required lengthy journeys (two to three years), nuclear propulsion systems are being contemplated. The crews of six to eight cosmonauts are projected to be all male with ages between 35 and 45 years.

Complicating the task of space exploration, however, is the reality of contemporary Soviet social and economic difficulties. Presently in a state of near-chaos, the country's financial future appears difficult to assess at best, and a disaster of indeterminate proportions at worse. In the midst of this difficult situation, as budgetary priorities pull at the Kremlin's purse strings from all directions, space stations and rocket-propelled boosters are becoming more and more difficult to justify. Accordingly, commercial aspects of the hardware are being proffered to Western customers with considerable regularity, and it is hoped that Western launch vehicle needs will take up much of the slack that now has developed in the Soviet launch capacity.

With a space program that has permitted cosmonauts to log thousands of hours in space, build and launch into orbit over two thousand satellites and at least eight space stations, and create a family of booster rockets that is second to none in the world in terms of lift capability and dependability, it will take more than these present domestic difficulties to hobble it. The Soviets are proud and strong and capable, and they will persevere ... As the great Tsiolkovsky noted during 1935, "Everything of which I speak is merely a feeble attempt to foresee the future of aviation, aeronautics, and rocketry. In one thing, I firmly believe that the Soviet Union will be first . . . ." ●

(right) *The* Nika-T *scientific research satellite is due to begin replacing the* Photon *satellite series during 1994, and will be used for weightless environment materials research. Capable of 120-day missions in a sun-synchronous orbit, it will carry recoverable payloads weighing up to 2,646 lb [1,200 kg].*

SUGGESTIONS FOR FURTHER READING

We recommend the following selected titles for further reading on matters concerning *Soviet Space:*

Baker, David *Conquest* (London, 1984).
Baker, David *The History of Manned Space Flight* (London, 1981).
Baker, David *The Rocket* (New York, 1978).
Caprara, Giovanni *Space Satellites* (New York, 1986).
Clark, Phillip *The Soviet Manned Space Program* (New York, 1988).
Froehlich, Walter *Apollo/Soyuz* (Washington, D.C., 1976).
Gatland, Kenneth *The Illustrated Encyclopedia of Space Technology* (New York, 1989).
Gunston, Bill *The Illustrated Encyclopedia Of The World's Rockets & Missiles* (New York, 1979).
Johnson, Nicholas *Soviet Military Strategy In Space* (New York, 1987).
Lebedev, Valentin *Diary Of A Cosmonaut: 211 Days In Space* (College Station, 1988).
Miller, Jay *OKB MiG, A History Of The Design Bureau And Its Aircraft* (Midland Counties, 1991).
Miscellaneous *The McGraw-Hill Encyclopedia of Space* (New York, 1967).
Newkirk, Dennis *Almanac of Soviet Manned Space Flight* (Houston, 1990).
Oberg, James *Red Star In Orbit* (New York, 1981).
Riabchikov, Yevgeny *Russians In Space* (New York, 1971).
Stoiko, Michael *Soviet Rocketry* (New York, 1970).
Wilson, Andrew *Solar System Log* (London, 1987).
Winter, Frank *Prelude To The Space Age, The Rocket Societies: 1924-1940* (Washington, D.C., 1983).

Index

A

Able   xviii
Aelita   103
Afghanistan   90
Agena   xviii
Akiyama   xix
Aktivny   103
Aldrin   xviii
Alexandrov   xix, 97
Allen   xix
Almaz   xix, 26, 103
Amateur Radio   27
Anders   xviii
Anikeyev   49
Anna 3   70
APEKS   103
Apogee   103
Apollo   xviii, 32, 64, 65, 67, 68, 96, 97
Apollo (ASTP)   xviii
Apollo-Soyuz   95
Apollo-Soyuz Test Project (ASTP)   95, 96, 97
Arcad (Arctic Auroral Density)   48
Archimedes crater   29
Armstrong   xviii, 68
Army Experimental Station Peenemunde   12
Artemev   11
Artyukhin   75
Astron/Granat   23
Atkov   xix
Austria   90
Aviavnito   10

B

Baikonur Cosmodrome   16, 21, 26, 49, 55, 57, 60, 75, 76, 78, 101, 102
Bake   xviii
Belka   20
Belyayev   xviii, 49, 60
Beregovoi   67
Berezovoi   xix, 80, 83
Bion   24, 26
Bluford   xix
BOR-4   101
Borman   xviii
Brand   xviii, xix, xx, 96
Brandenstein   xix
Britain   90
BST-1M telescope   77
Buchli   xix
Bulgaria   78, 90, 96
Buran   99, 100, 101, 102, 103
Bykovsky   xviii, 49, 56

C

Carr   xviii
Cernan   xviii
Chaffee   xviii
Challenger   xix
Chaplygin   11
Chibis   74
Co-Orbital Anti-Satellite (ASAT)   28
Cold War   xx, 31
Collins   xviii
Columbia   xix
Conrad   xviii
Cooper   xviii
cosmodromes   16
cosmonauts   49
Cosmos   21, 26
Covey   xix
Crippen   xix
Cuba   78, 96
Cunningham   xviii
Czechoslovakia   78, 81, 96

D

D-module–Doosnashcheniye ("additional equipment")   90
De la terre a' la lune   1
Design Bureau No.7   10
Diary Of A Cosmonaut: 211 Days In Space   83
Dobrovolsky   xviii, 72
docking module   96
Dornberger   13
Duke   xviii
Dunbar   xix
Dushkin   2, 11
Dzhanibekov   xix, 54, 73, 78

E

Echo   xviii
Egypt   96
Eisele   xviii
Ekran   27, 102
Electronics Intelligence Ocean Reconnaissance Satellites (EORSAT)   28
Elektron   23

Endeavour   xix
Energiya   xix, 90, 98, 99, 101, 102
Energiya/Buran   100
Enterprise   xix
EO (Ekspeditsya osnovnoi /"Principal Expedition")   77
EORSAT   28
EP (Ekspeditsya poseshchenya /"Visiting Expedition")   77
Evans   xviii
Experimental Rocket Engine #1   7
Explorer   xviii, 20, 21
extra-vehicular activity (EVA)   56, 67, 92

F

Fabian   xix
Faith   xviii
Federation Aeronautique International (FAI)   80
Fedorov   11, 49
FEK-7 photoemulsion camera   70
Feoktistov   xviii, 16, 60
Filatyev   49
Filipchenko   67
Fisher   xix
Fractional Orbital Bombardment System (FOBS)   28
France   90
Freedom   xviii, 90
Friendship   xviii
Furrer   xix

G

Gagarin   xviii, xx, 24, 49, 51, 55, 57, 59, 68, 70, 83
Gagarin Training Center   55
Galileo   xix
Gamma   103
Gardner   xix
Garneau   xix
Garriott   xix
Gas Dynamics Laboratory   6
GDL   6, 7, 15
Gemini   xviii, 76
German Democratic Republic   78
German rocketry   12
Germany   96
Giacobini   xix
Gibson   xviii, xix
GIRD   2, 4, 5, 7, 8, 15
glasnost   63
GLAVKOSMOS   xix
Glazkov   75
Glenn   xviii
Global Electronic Intelligence   28
Global Navigation Satellite System   27
GLONASS   27, 28
Glushko   6, 11, 99
Goddard   1, 6, 12
Gorbatko   49, 67, 75
Gordon   xviii
Gorizont   27, 102
Granat   xix, 22, 23, 102
Grave   6, 11
Grechko   xix, 75, 78
Grissom   xviii
GRO – Gamma Ray Observatory   xix
Gröttrup   13
Group for the Study of Reactive Motion   6
Gruppa po izucheniyu reaktivnogo dvizhenia   6
Gryaznov   11
Gubanov   102
Gubarev   xix, 75
Gurevich   99

H

Haise   xviii
Halley's Comet   xix, 39, 40
Hart   xix
Hartsfield   xix
Harz Mountains   13
Hauck   xix
Heavy Kosmos   76, 77, 99
Hermaszewski   81
Hilmers   xix
HST – Hubble Space Telescope   xix
Hungary   96

I

I-module–("International Ecological Research")   90
ICBM   14, 15, 17, 50
Ikarus ("space motorcycle")   xix, 92
Il'yin   10
India   96
Informator   xix
Institute for Aerodynamic and Hydrodynamic Research   14
Interball   103
intercontinental ballistic missiles   14, 15, 17
intercontinental ballistic rocket   17
Interkosmos   21, 78, 81, 96, 102
International Cometary Explorer   xix

International Geophysical Year   17
Intersputnik   27, 96
Iran   96
Irwin   xviii
Isayev   11
Iskra   27, 83
Ivanchenkov   xix
Izhevskoye   1

J

Japan   90, 96
Jarvis   xix
Jupiter   xviii

K

Kaliningrad   49, 75
Kaluga   1
Kapustin Yar   16
Kapustin Yar Cosmodrome   101
Karpov   49
KATE-140 wide-angle, stereographic/topographical camera   77
Katyusha   10
Kazakhstan   16
Kazbek   91
KB-7   10
Kennedy   xviii, 63
Khodynskoye Field   51, 55
Khrunov   xviii, 49, 67
Khrushchev   17, 21
Kibalchich   11
Kiev   6
Kizim   xix, 83
Kleimenov   8, 10
Klimuk   75
Kolbasicha   2
Komarov   xviii, 60, 67
Kondratyuk   11
Korabl-Sputnik   xviii, 51
Korneyev   2, 11
Korolev   2, 7, 8, 9, 10, 11, 12, 14, 15, 17, 20, 49, 56, 63, 68, 99, 101
Koronas   103
Kosmograd ("Spacetown")   90
Kosmos   xviii, xix, 21, 26, 27, 28, 62, 63, 65, 72, 76, 95, 101, 102
KOSPAS   96
Kostikov   10, 11
Kosygin   xviii, 95
Kovalenok   xix, 78, 83
Kristall   xix, 84, 90
Kubasov   xviii, xx, 67, 71, 96
Kvant   xix, 83, 84, 86, 87

L

Laika   xviii, 17, 20
Langemak   10, 11
Lapirov-Skobolo   6, 11
Laveikin   xix, 87
Lazarev   xviii
LDEF – Long Duration Exposure Facility   xix
Lebedev   xix, 80, 83, 87
Leestma   xix
LenGIRD   6, 7
Leningrad Institute of Communication Engineers   6
Lenoir   xix
Leonov   xviii, 49, 60, 63, 73, 96, 97
Levchenko   xix, 87
Lichtenberg   xix
Lomonsov   103
Lounge   xix
Lovell   xviii
Luna   xviii, 29, 30, 31, 32, 33, 35
Lunar Orbiter   xviii
Lunokhod   xviii, 34, 35
Lyakhov   xix, 81

M

Magellan   xix
Makarov   xviii, xix, 78
Manarov   xix, 87
Manhattan Island   26
Manned Maneuvering Unit   xix
Mariner   xviii
Mars   xviii, 44, 45
Mattingly   xviii, xix
McAuliffe   xix
McBride   xix
McCandless   xix
McDivitt   xviii
McNair   xix
Medilab   90
Merbold   xix
Merkulov   11
Meshchersky   11
Messerschmid   xix
Meteo   14
Meteor   27, 102

Meteor-Priroda   26, 27
MIG   99
Mikoyan   99
Mir   xix, 72, 83, 84, 86, 87, 90, 95, 102
MKF-6M earth resources camera   77
Molniya   27, 38, 78, 102
Mongolia   96
Moscow   1
Moscow Aviation Engine Plant   10
Moscow GIRD   2
MosGIRD   6, 7
Moshkin   11
MR-1   14

N

N-1   64, 65, 101
N.E. Zhukovsky Air Force Academy   6
Nadezhda   27, 96, 102
Nagel   xix
NASA   95
Nataliya   103
Navstar-GPS   27
Nelson   xix
Nelyubov   49
Neytron   103
Nika-T   23, 104
Nikitin   49
Nikolayev   xviii, 49, 55, 67
Nixon   xviii, 95

O

O-module–("Optical")   90
Oasis 1   70
Oberth   1, 6
Ockels   xix
Onizuka   xix
Optynyy raketnyy motor   7
OR-1   7
OR-2   7
Oreol   48
Orion   70
Orlan   92
ORM-1   7
Osoaviakhim   7, 10
Overmeyer   xix

P

P.K.R.D.D.   14
Parker   xix
Patsayev   xviii, 72
Peenemunde   12, 13
Penguin suit   70, 74
Perelman   1, 11
perestroika   63
Petrapovlovsky   11
Phobos   xix, 44, 47, 103
Photon   23, 24, 26, 73, 102, 104
Pioneer   xviii, xix
Pioneer-Venus   xix
Plesetsk   16, 57, 102
Plesetsk Cosmodrome   27
Pobedonostev   11, 14
Pogue   xviii
Poland   96
Polish   81
Polyarny   2, 7, 11
Polyot   21
Popov   xix, 83
Popovich   49, 55, 75
Pravda   17
Pravitel'stvennaya komissiya po raketam dalnego deistviya   14
Priroda   90
Prognoz   23, 102
Progress   xix, 75, 76, 77, 78, 81, 86, 87, 90, 102
Progress-M   xix
Proton   xviii, 23, 40, 60, 64, 69, 70, 72, 75, 77, 82, 83, 87, 102

R

R-06   4, 7
R-07   8
R-14   14
R-7   15, 16, 17, 21, 49, 50, 56
Radar Ocean Reconnaissance Satellites (RORSAT)   28
Radioastron   103
Raduga   27, 102
Rafikov   49
Ranger   xviii
RATO   10
Razumov   11
RD-253   60, 72
RD-301   66, 67
reactive motion law   2
Reaktivni nauchno-issledovatel'kii institut   8
Regatta   103
Relikt   103
Remek   xix, 81

Resnik   xix
Resurs   28, 102
Resurs-O   26
Resurs-F   26, 28
Revolutionary Military Council's Department of Armaments   10
Ride   xix
RNII   8, 10, 14, 15
Romaneko   xix
Romanenko   xix, 78, 87, 90
Romania   96
RORSAT   28
RP-1   7, 8
Rukavishnikov   72
Rynin   1, 6, 11
Ryumin   xix, 78, 81, 83

S

Salikov   11
Salyut   xviii, xix, 54, 68, 70, 72, 73, 75, 76, 78, 80, 82, 83, 87, 95
Sandal   17
Sapwood   17
SARSAT   96
SAS   103
Saturn   96
Saturn V   64
Saturn V booster   xviii
Savinykh   xix, 83
Savitskaya   xix
Schirra   xviii
Schmitt   xviii
Scientific Research Institute of Jet Propulsion   8
Scobee   xix
Scott   xviii
Scully-Power   xix
Serebrov   xix, 91
Sevastyanov   xviii, 67, 75
Shatalov   xviii, 67, 72
Shaw   xix
Shonin   xviii, 49, 67
Shtern   11
Shyster   17
Skylab   xviii, 72
SL-1   15, 21
SL-13   69
Slayton   xviii, xx, 96
Smith   xix
Society for Assisting Defense and Aviation and Chemical Construction in the USSR   7
Society for the Study of Interplanetary Communication   6
Society for the Study of Interplanetary Travel   6
Sokol   71, 91
solar system   103
Solovyov   xix, 83
Soviet nuclear weapons   15
Soviet shuttle   xix
Soviet Union's Academy of Sciences   95
Soyuz   xviii, xix, 21, 50, 51, 54, 55, 56, 63, 64, 65, 67, 68, 70, 72, 73, 75, 76, 77, 78, 81, 83, 86, 87, 88, 90, 91, 95, 96, 97, 102
Soyuz 19 (ASTP)   71
Soyuz-Apollo   xx
Soyuz-Salyut   82
space motorcycle   xix
space shuttle   99
Spain   90
Spektr   90, 103
Spiral   101
Splav 02   73
Sputnik   xviii, 17, 19, 20, 21, 37
SS-6 Sapwood   15
Stafford   xviii, xx, 96, 97
Stalin   6, 10, 17
Stankyavichyus   101
Star City   55, 70
State Commission for the Study of the Problems of Long-Range Rockets - (P.K.R.D.D.)   14
Stechkin   11
Stewart   xix
Strekalov   xix
Strelka   20
STS   xix
Sullivan   xix
Swigert   xviii
Syria   96

T

T-module–("Technological")   90
Tamayo   xix
Telstar   xviii
Tereshkova   xviii, 56
Thagard   xix
The Great Patriotic War   10
Thornton   xix
Tikhomirov   6, 11
Tiros   xviii
Titan   xix
Titov   xviii, xix, 49, 51, 55, 87
Truly   xix

TsAGI   14
Tsander   6, 7, 8, 11, 15
Tsikada   27, 96
Tsiklon   102
Tsiolkovsky   xxiii, 1, 11, 15, 49, 104
Tukhachevsky   6, 10
Tyuratam   16

U

U-2   16
Ugolek   62, 63
Ulysses   xix

V

V-2 (Vengeance Weapon 2)   12, 13, 14, 15, 16
V-2-A   14
Van Allen radiation belts   23
Van Hoften   xix
Vandenberg A.F.B.   xix
Vanguard   xviii
Vasyutin   xix
Vega   xix, 40, 42, 43, 103
Vegas   39
Venera   xviii, xix, 36, 37, 38, 40
Veneras   39
Venus   44
Vernadsky Institute of Geochemistry and Analytical   30
Verne   1, 6
Vetchinkin   6, 11
Veterok   62, 63
Vietnam   96
Viking   xviii
Volk   xix, 101
Volkov   xviii, xix, 67, 72
Volynov   xviii, 49, 67, 73
von Braun   13
Voskhod   xviii, 56, 60, 63, 65
Voskhod 2   xx
Vostok   xviii, 23, 24, 49, 50, 51, 52, 55, 56, 59, 65, 68, 78, 102
Vostok Zh   56
Voyager   xix
Vzor   49

W

Walker   xix
Wells   6
White   xviii
Wolgast   12
Worden   xviii

Y

Yar   57, 102
Yastreb   71
Yegorov   xviii, 60
Yeliseyev   xviii, 67, 72
Young   xviii, xix

Z

Zaikin   49
Zenit   102
Zhukovsky   11
Zinner   xix
Zond   xviii, 31, 37, 44, 65
Zvezdny Gorodok   55

GLOSSARY

ASAT—anti-satellite.

ASTP—*Apollo/Soyuz* Test Project.

astrophysical—relating to the physics of astronomy and the celestial bodies.

Baikonur—one of the major Soviet launch complexes.

barycentric—an orbit of a common center of mass of one or more bodies.

booster—the large first stage of a multi-stage rocket.

cosmos—an ordered, harmonious universe.

EO—*Ekspeditsya osnovnoi* ("Principal Expedition").

EP—*Ekspeditsya poseshchenvy* ("Visiting Expedition").

EVA—Extra-Vehicular Activity.

FOBS—Fractional Orbital Bombardment System.

frustrum—a truncated cone.

GDL—Gas Dynamics Laboratory.

geostationary—term used to describe a satellite that remains in a fixed position over a given equatorial spot on the Earth's surface.

GIRD—*Gruppa po izucheniyu reaktivnogo dvizhenia* ("Group for the Study of Reactive Motion").

GLAVKOSMOS—"The Central Administration of Space Technology Development and Use for the National Economy and Science"

GLONASS—Global Navigation Satellite System.

GPS—Global Positioning System.

ICBM—Intercontinental Ballistic Missile.

Kapustin Yar—one of the major Soviet launch complexes.

kg—kilogram/kilograms [equivalent to 2.205 lb].

km—kilometer/kilometers [equivalent to 0.6214 mi].

km/h—kilometers per hour.

Kosmograd—"Spacetown."

KOSPAS (or *COSPAS*)—the Soviet segment of the international search and rescue system; known as SARSAT in the U.S.

lb—pound/pounds

LenGIRD—*Leningradskaya gruppa po izucheniyu reaktivnogo dvizhenia* ("Leningrad Group for the Study of Reactive Motion").

MosGIRD—Moscow's *Gruppa po izucheniyu reaktivnogo dvizhenia* ("Group for the Study of Reactive Motion").

mph—miles per hour.

OIMS—All-Union Society to Study Interplanetary Communications.

ORM—*Optynyy raketnyy motor* ("Experimental Rocket Engine").

Osoaviakhim—"Society for Assisting Defense and Aviation and Chemical Construction in the U.S.S.R."

payload—the scientific research package or cosmonaut cargo carried aloft by a rocket.

Peenemunde—German rocket research facility born just prior to World War II and captured by the Soviets afterward.

PKRDD—*Pravitel'stevennaya komissiya po raketam dalnego deistviya* ("State Commission for the Study of the Problems of Long-Range Rockets").

Plesetsk—one of the major Soviet launch complexes.

RATO—rocket-assisted take-off.

RNII—*Reaktivni nauchno-issledovatel'kii institut* ("Scientific Research Institute").

salvo—to discharge rockets, missiles, or artillery in rapid succession.

SARSAT—Search and Rescue Satellite.

satellite—a small manmade or natural body which orbits a larger one under the influence of gravitational pull.

subsonic—a velocity less than that of sound; (the speed of sound is approximately 760 mph at sea level and 670 mph at altitudes of 35,000 ft and above).

supersonic—a velocity greater than the speed of sound.

TsAGI—"Central Aerodynamic and Hydrodynamics Institute."

TsBIRP—"Central Bureau for the Study of the Problems of Rockets."

Tyuratam—A railway junction town east of the Aral Sea. The so-called Baikonur Cosmodrome is located nearby.

V-2—*Vergeltungswaffe-zwei* ("Vengeance Weapon Two"); a German rocket with warhead utilized with considerable effect during the last seven months of World War II.

Venusian—of or pertaining to the planet Venus.

METRIC CONVERSION CHART

Multiply	By	To Obtain
Feet	0.3048	Meters
Meters	3.281	Feet
Kilometers	3281	Feet
Kilometers	0.6214	Statue miles
Statue miles	1.609	Kilometers
Nautical miles	1.852	Kilometers
Nautical miles	1.1508	Statue miles
Statue miles	0.86898	Nautical miles
Pounds	0.4536	Kilograms
Kilograms	2.205	Pounds
Feet/second	0.3048	Meters/second
Meters/second	3.281	Feet/second
Metric ton	1.102	U.S. ton
U.S. ton	0.907	Metric ton